T0181266

Mind and Nature
Essays on Time and Subjectivity

Mind and Nature

Essays on Time and Subjectivity

JASON W. BROWN

Clinical Professor, Department of Neurology
New York University Medical Center

W

WHURR PUBLISHERS
LONDON AND PHILADELPHIA

© 2000 Whurr Publishers
First published 2000 by
Whurr Publishers Ltd
19b Compton Terrace, London N1 2UN, England and
325 Chestnut Street, Philadelphia PA 1906, USA

British Library Cataloguing in Publication Data
A catalogue record for this book is available from the
British Library.

ISBN: 1 86156 148 2

Contents

Tumult and peace, the darkness and the light –
Were all like workings of one mind, the features
Of the same face, blossoms upon one tree;
Characters of the great Apocalypse,
The types and symbols of Eternity,
Of first, and last, and midst, and without end.

Wordsworth

Preface

This book picks up where the last one left off, with what was left unsaid or unfinished, or with ideas that deserved further explication, as if I could never quite say enough about a topic, or perhaps, as a fact of psychological interest, that between any two sentences there is another book just waiting to be written. Like my other books, all of which followed an early period of clinical studies, it is driven by a single theoretical perspective, with the original data being revived, not as facts to be referred to or justifications, but as patterns of thought that guide the intuitions along, so that a reader hungry for evidence, or for the research or clinical material on which the theory is constructed, will have to retrace the path taken by the author himself, if he or she wishes to fully understand the coherence of statements in an exposition that is a diachronic evolution of thought, in the same way that the theory itself is evolutionary.

At the very outset, I want to stake out, in simple terms, the position taken in this book. The claim is that Cartesian dualism is still very much the problem to be addressed. Those who postulate a reduction of mind to brain, or an elimination of mental events in the name of future neuroscience have, it seems to me, a laziness of thought, or expediency, that makes them gloss too readily over issues of great depth that have troubled the best philosophical minds. On the other hand, the nowadays rare individual who takes an idealist position, and I was myself long in this camp, needs to find a path back into the physical world which, I think, is helped by considering the relevance for philosophy of clinical phenomena, where the physical and mental degrade as a unit. In this respect, this book records a shift in my own thought from idealism to process monism, in which the physical and mental are viewed as manifestations at successive stages in the evolution of a single process.

How could there not be a continuum from the non-cognitive to consciousness? This continuum, however, need not run from a low level *physical* series to one of increasing mentality. If we cannot describe the passage from physical nature to mind in terms other than purely physical, or purely mental, that is, from one side or the other of the Cartesian impasse, our theories of mind and nature must be defective. Matter does not become mind at some point of intricacy but anticipates mind at every point. There is no gap or Rubicon beyond which mind appears. A continuum requires the physical and mental to evolve together into progressively more complex ensembles. This scenario is not pan-psychic, but it does entail that the (pre)conditions for subjectivity – temporal becoming and categorical existence – obtain at every level and in every actuality, so that mind develops, so to say, from within by an expansion of the becoming of basic entities. This book is a search for the laws of that evolution, not by the more typical strategy of presuming that consciousness is a compound of the data of physics, or that the lower the level of explanation, the more powerful, but by tracing the categories of experience from the cognitive down to the foundational process of nature and then, by an adjustment of physics, recovering an adequate theory of the mental from its basic ingredients. As Heraclitus said, the road up and the road down are the same.

What is meant by a category of experience? The consciousness of an object engages the mental state in an act of cognition that includes the object as its 'objective' segment. This is itself an object with a complex structure. The correlate of the perceived object 'out there' in the world, in relation to which the perception appears like a screen, is a segment in a phase-transition in the brain state. The immediate correlate of a tree in my conscious perception is the brain activity that generates the image of self and world of which the tree is one part. The perceived object, i.e. the 'objective' portion of my mental state, is the nominal referent of the act of cognition, but in order to achieve that object many phases in the mind/brain state intervene, and each could be said to correspond with some aspect of the external world. I think it is preferable to hold that these phases correspond with phases in brain process that, collectively, give a model of externality.

What externality is being modelled? The object 'behind' the perceptual object, the real or noumenal object, the object that is registered by neurosensory data, can be taken to exist as another sort of entity, more or less complex than the mind/brain state that perceives it. This object has its own dynamic structure, whether an atom, a tree or another person, with phase transitions of varying complexity. Of the totality of actualities that constitute the world, including the minds that actualize in that totality, each one has a transitional and a categorical 'structure'. The categorical existence of the phase-transition of a basic object constitutes its incipient subjectivity, its 'in itselfness', independent of the mind that happens to perceive it.

A percipient mind actualizes a self, acts, memories, ideas, images, objects. In the course of an actualization, the self is an antecedent phase, the object a consequent phase in the same process. The entire fabric of an act of cognition, from self to world, is a single process. The break between mind and world, the disjunction of the self from its objects, is illusory. Every cognition is molded by the 'out there' to generate a more or less accurate model of reality and constrained by patterns of neural process that tend to be repeated. A tree is revived every moment, it becomes what it is, it does not just persist, it renews itself, and each renewal is slightly different than the prior occasion. This difference is the basis for novelty. The novelty in the renewal is the basis of change. This is also the source of novelty in an act of cognition. The papers in this collection are all guided by this metaphysical presupposition.

Part I

The initial section of this book deals with accounts of 'becoming' in the two major systems of philosophical thought that have affinities with microgenetic theory, the process philosophy of Whitehead and the Buddhist concept of the arising and perishing of the moment. These are both process theories that provide an alternative to the hegemony of contemporary substantialism. The doctrine of substance has a long history in science and philosophy. The pervasiveness of the doctrine, and the intuitions that support it, reflect the tendency of the human mind to both form categories and analyse them. If metaphysical presuppositions are underwritten by an unconscious cognitive bias, what is that bias, and what are its origins?

Substance theory is inevitable. Perception creates objects and a subject to enjoy them. The idea of substance begins with the first partition of a thing from its context, whether subject and object, an individual from its environment and history, or an atom from the duration of its cycles. Indeed, the first appearance of the discernible out of the invisible is already a commitment, a discrimination. Boundaries are established that mark off separate things. However, for process philosophy and Buddhism, a concrete entity is an actualization, an outcome, not a demarcation or a starting point. In this process, indeed because it is a process, time and change play a central role. The problem for process thought is not so much the transition from one solid to another, though how this occurs is far from clear, but the more fundamental question of how the 'solids', the phenomenal stabilities or the concrete actualities, are created in the first place. This is not as great a difficulty for substance theory, which assumes self-standing causal entities. In the critique of the later Buddhists, they have *svabhava*, self-nature, that is, they are aggregates of discrete events or synchronic compounds of solids that are still more primitive.

Whitehead's response to this problem is the idea of a concrescence, a *prehension*, a process of *unification*, that is bounded by an arising and a perishing, a becoming of time and entities out of, and relapsing into, the timelessness of eternal ideas, while in Buddhism, there is an unspecified *construction* of phenomenal entities out of the flux of the Absolute that, for oneness, in enlightenment, is stripped away to expose an underlying relationality that is fundamental. In a sense, process philosophy fills in the gap between the concrete world and the Absolute that Buddhism leaves unexplained. Yet, some schools of early Buddhism postulate elemental substances in the Absolute, and even some theorists of process, including Whitehead, concede to substantialism the changelessness of entities such as mathematical concepts, eternal or Platonic ideas, Kant's noumenal self, which, being out of time, thus timeless, constitute a kind of radicalization of time elimination.

That a theory of the mind/brain which developed on the observation of clinical disorders has led quite independently to an account of process that maps to these two great metaphysical systems is a forceful argument for the correctness of those insights on which the theories are based. Further, the concordance of the philosophy with patterns of behavior, inferred from pathology, suggests that the laws of mind and nature are reciprocally discoverable, that is, that mind is an outcome of the process of nature and, similarly, that metaphysical ideas are products of the form-creating activity of the mind. This has been argued by, among others, Eddington, who claimed that the fundamental laws of nature could be derived from the laws of human mentation. He wrote, 'we must regard the feeling of "becoming" as (in some respects at least) a true mental insight into the physical condition which determines it.' Dean Inge put it more poetically: 'the same power which slumbers in the stone and dreams in the flower, awakens in the human soul'.

A theory of becoming is retrospective, it is the creation of the present out of the past, or the revival of the past in every actual occasion. In this manner, becoming is tied to evolutionary and genetic concepts. An evolutionary perspective is essential if thought is to be conceived as a species of process in nature, that is, if the laws of physical nature are held to apply to cognition. In process metapsychology, an act of cognition is a momentary actualization, a becoming, that replicates patterns of evolutionary growth. These patterns, like the deeper process of change they represent, are concealed from observation. That is why they are best studied in cases, such as those of mental disorder, where the submerged is suddenly thrown into view. The study of such cases reveals the lines of dissolution, and by implication the formative dynamic, that recapture and sustain the growth trends of non-cognitive nature into mental process. Thought and perception are modes of growth, and growth is a mode of change. The uniformity of change in cognitive and non-cognitive systems, and the correspondence of this process with evolutionary patterns of growth, provides a basis for a metapsychology of organic and inorganic nature.

Part II

The two papers on Freud's metapsychology bring into relief some of the philosophical problems attendant to a description of the laws of mentation in relation to a source of clinical data other than that of brain pathology. The Freudian corpus constitutes the sole metapsychology, apart from that of microgenesis, that is based on clinical studies. In my opinion there are grave difficulties with the foundational concepts of psychoanalysis, though certain features of the topographic and genetic models are relevant to process thought, as in the objectification of the conscious present out of the personal past. Moreover, though psycho-analysis is concerned more with unconscious than conscious thought, the implicit progression from the unconscious to consciousness contains the seed of an actualization concept that, at least in the hands of Schilder and Rappaport, has some resemblance to a cognitive becoming. On the other hand, the two most important papers on the metapsychology, the *Project*, and the *Unconscious*, do not provide an authentic account of *process* in mind or brain, since Freud relied almost completely on drive-energetics to introduce an *ad hoc* dynamic that could activate the benign surfaces of inert traces.

Consciousness is, so to say, on everyone's mind, like a crown of laurels, it is the holy grail of current philosophizing. The claim that consciousness is the most complex of all known phenomena is easy to make, hard to prove. Certainly, there is complexity in the discussion of the topic, but is consciousness equally complex? I would, in fact, argue the reverse, that consciousness is exceedingly simple! Proclus said, only the highest and lowest are simple, while all between is complex. What could be more simple than the timelessness of a virtual memory in which the mental state arises, or the timelessness of a virtual duration, the conscious present, in which it perishes, while between the two (essentially the entire field of psychology) one finds the multiplicity of temporal facts and attachments of everyday life, the simplicity of an eternity on either side of complex temporal data, a self-creation out of eternity that is, as Goethe wrote, an ever-creating vocation.

The simplicity of consciousness is linked to the nature of categories, the topic of the ensuing paper. Natural categories condition theory to reflexively assume that all entities are relationless, or that temporal and spatial connectivities can be pasted to object boundaries. Categorical frames of thought become fixed entities dividing permanence and flux. The flux is an abstraction *within* the category, the category, an abstraction *over* the flux; substance is category all the way through. By an alignment of thinking with the natural direction of analytic thought, beginning with those of the 'highest' grade – the boundedness of the self, its separation from other objects, introspection, self-scrutiny – the division of categories proceeds by an increasing partition *within* the self down to the unseen particles of final origination that are nested in the Absolute.

That a category is a mental structure does not mean it is a product of mind alone. The continuity of the categories of mind with those of nature is the metaphysical ground of autonomy, which is a stage in the process of category formation. A substance, an object, a self, are all categorical objects. The boundaries of entities are fuzzy at the extremes of scale. Consciousness and quanta are categories rather than things. In between, in the range of mid-sized objects such as chairs and trees, entities no longer seem to be categories; instead, they appear to be concrete and well delineated. The categorical is above us and below us, but invisible – objectified – in the middle range where life passes.

In this middle ground the perception of stability is overpowering; we cut events out of continua and objects out of events. Even a point-instant, in Buddhism or in physics, subsumes a temporal thickness without temporal parts that is derived from fundamental temporal extensibility and is spread out over the cycle of its continuation (becoming). Timothy Sprigge has written that continuants are 'concrete universals' in the relations of the temporal series that constitutes their becoming. The duration and phases of a becoming are inter-dependent. The process lays down the object, the category gives it the stability it requires to be what it is. Becoming is the key. An actuality never endures, concretely, it is never fully present, but perishes as it actualizes. Referring to this immediate pastness of actualities and the retrospective quality of mentation, Charles Hartshorne wrote that the paradigm for cognition is history, not mathematics.

Part III

The final section extends microgenetic theory to the nature of the self-concept and its relation to culture, value and aesthetics. In some ways, this is a search for the moral implications of the metapsychology. Consider Whitehead's remark, 'I sometimes think that all modern immorality comes from the Aristotelian doctrine of substance'. While this gives too much to philosophy and too little to human nature, what is the connection between substance and immorality? A self that observes its own acts is the first step in a moral decision. The next is a choice *between* acts. The immoral is not spontaneous. An agent must *evaluate* an act in relation to the objects of its needs or desires. Agency is autonomy in relation to choice. Choice presupposes agency but is not obligated by it. Nor does the choosing entail *real choices*, rather, the *availability in the mind of options*, or the awareness of possibility.

Morality is a judgment based on choices, but it is conditioned on a projection of value to the other, really, an absorption of the other in the self's own valuation, which is the antithesis of autonomy. The distribution of values inclines the agent to that option on which the moral judgment rests. Ultimately, these are the self's own attachments. In the maturation of a personality, ideally, the field of value widens from family to friends to nation, to the point where, as Hartshorne wrote, the individual is a 'trustee for living things in its part of the cosmos'. Emmanuel Levinas wrote, 'I am responsible for the other without waiting for reciprocity'. To love, to take someone *into* your heart, to *share* your feelings with another, to be vulnerable to loss, is to subsume another in your boundedness. Fear and anger are also valuations. The other is not a neutral entity; it threatens to violate the autonomy of the self, its impunity. Unlike the sought-after, the willing embrace of love, the other penetrates the self as an intruder. This widening of the personality is a reclaiming of antecedent phases of animistic thought usually traversed in the surge to objectivity. In this respect, the other becomes an artwork, an aesthetic object, a creation of feeling prior to boundaries, separation, individuation. In loving, in empathy, the subjectivity of primitive thought becomes the shared experience of a common humanity.

A phase of magical thinking is engaged in every act of cognition. The primitive in thought is the ground in which the psyche individuates. Even the idea of substance is a remnant of animism, where the causal power in objects is the agentive aspect of the psychic deprived of its freedom. The psyche is bathed in otherness as it strives for independence. Autonomy is set against this background. Dorothy Emmet has written that the primitive does not project his self-awareness into the world but 'starts from a sense of the continuity of his functions and activities with those of an environing world . . . [and] as conscious thought develops these vague surrounding potencies are endowed with form and thereby "objectified", and at the same time man comes to a consciousness of himself as a distinct subject over against them.'

Feelings are tokens of subjectivity. They anticipate consciousness as covert motivations, go beyond the limits of privacy and deposit in external objects. Subjectivity is constantly being lost in actuality. We can retain the subjective by extending it into nature, as in primitive thought, in art or in religion, where objects are imbued with psychic qualities, or by drawing those objects within us, to their origins, to the interioricity that is the basis of the aesthetic response. The renunciation of psychic individuality to the eternity and immutability of the Christian God, if not that of Whitehead, differs from the entrance to the Buddhist nirvana, which is an unveiling of the phenomenal for the sheer becoming of the Absolute. This distinction pivots on an axis of feeling in a transcendent immersion. Does one follow a stream of feeling from object valuation through its underpinnings, in feeling, to the instinctual unconscious, primal feeling or will, or is feeling tone a subjective distraction in the pursuit of naked objects back into the void? If nature is suffused with feeling, a dissolution of the temporal would leave total absorption in a *felt* eternal present.

The contextuality that remains behind in the attainment of autonomy softens isolation with humanity. It is not the autonomy but the residual valuation of the self, its goals, that determine the locus of an action within the self-concept. The autonomous self must find its way back into the world. Devotion is a doorway to otherness. We are helped to recognize all selves in the One, a uniformity that becomes a diversity when the potential in unity inclines to some path of actualization. The futurity of potential, its implicit aim, is its lack of homogeneity, even if the specificity of future parts has not yet been established. Like notes in the score of life, we actualize, as Tagore said, in 'the music of the great I AM'.

In great souls, the self disowns autonomy and surrenders to the mind of deity. The rediscovery of oneness proceeds to a unity of the corporeal in the sea of being. Mind, the vehicle of this return, is shed once unity is achieved, for the categories of nature, continuous with mind, are bridges to their own awareness. We are One in matter and in Spirit, which is the life of matter in change. The great soul teaches us the oneness that was our birthplace, when we first emerged, from unity, like separate leaves on the tree of life. We know the unknowable because it is within us.

Jason W. Brown
Fontareches, France
October 1999

Acknowledgements

I am grateful to the following editors and journals for permission to reprint certain of my papers in this volume: chapter 1, Barry Whitney (*Process Studies* 27: 79–92, 1998); chapter 2, Roger Ames (*Philosophy East and West* 49: 261–277, 1999); chapter 3, Harry Whitaker (*Brain and Cognition* 38: 234–245, 1998); chapter 4, Robert Bilder (*Annals of the New York Academy of Sciences* 843: 91–106, 1998); chapter 5 was given as an invited lecture at a symposium of the American Psychological Association, Boston, August, 1999; chapter 7 was given as an invited lecture at a conference on Das Selbstkonzept und seine neurobiologischen Grundlagen, Bonn 1998, and published by Gene Brody (*Journal of Nervous and Mental Disease* 187: 131–141, 1999); chapter 8 is based on an invited lecture at a conference on Subjectivity in the Humanities, SUNY at Stony Brook, March 1998; chapter 9, Keith Sutherland (*Journal of Consciousness Studies* 6 (6/7): 144–161, 1999).

Part I
Metapsychology

Chapter 1
Foundations of Cognitive Metaphysics

Philosophical materialism enjoys a large body of empirical support in cognitive psychology precisely because the agenda of the psychology is motivated by the very concepts the philosophy endeavors to explain. This is not the case in process thought where the genetic concepts that underlie evolutionary and developmental theories have had scant impact on process studies and there is little or no interpenetration of philosophical analysis with psychological research. This is not a promising state of affairs.

The purpose of this chapter is to explore some of the implications for process philosophy of a new approach to brain psychology and the dynamics of the mental state – microgenetic theory – that has developed out of the study of symptoms in neurological cases. This approach has much in common with process metaphysics, especially in the concept of time, change and the actualization (becoming) over phases in the brain in the momentary development of a cognition.

Microgenesis and Process Theory

Microgenesis refers to the actualization (*Aktualgenese*) of a cognition over 'layers' in mind and brain that retrace growth patterns in phyloontogeny.[1] The recapitulation that is the cornerstone of historical theory is a repetition of the antecedents of a behavior that phyletic or ontogenetic process lays down. In its simple form, the theory held that a cognition develops over anatomical stages in the brain. These stages are entrained at successive phases and are aligned in a sequence that reflects evolutionary growth. This was the shape of early recapitulation theory as applied to brain and behavior. There were some who postulated a retracing of archaic repertoires that remained embedded in the final behavior, for example Paul MacLean's notion of a reptilian and protomammalian brain within the mature human brain.[2]

Gradually, it became clear that it is not the stages or the behaviors that are reproduced but the configural properties of the process through

3

which they actualize, that is, the process is revived, not the actual elements into which it deposits. Moreover, the earlier concept of a collapse of the millions of years of phylogeny, or the lifespan of ontogeny, into the milliseconds of a cognition, or the idea of a process that continued over evolutionary, lifespan and cognitive durations was replaced by the concept of an iteration of a single process or pattern that binds together the different time frames.

More precisely, the duration of phyletic or ontogenetic process is not the evolutionary (maturational) history of a species (organism); the former is more accurately the sum of its ontogenies. Evolution is a population dynamic, ontogeny the life story of an individual. From the individual standpoint, evolution is the antecedent line of all prior ontogenies for that organism. Thus, the question, what exactly is an ontogeny? The conventional view is of a process that extends over the life-span. But there is a way of regarding ontogeny as a moment of growth that is cyclically revisited. What is the lifespan if not a temporal aggregate that is woven by the mind into a seamless thread from the series of discrete momentary actualities.

If this is the proper way to interpret ontogeny, the duration we are seeking would not extend from infancy to senescence; ontogeny is not the longevity – the growth and decay – of the organism from birth to death. Rather, the duration of an ontogeny lies in the covert process that deposits the organism each moment and at every phase in its life cycle. In this way of thinking, the momentary actualization of the organism, its becoming, is the fundamental note from which the melody of development is composed.

Every becoming of the mental state (microgeny) creates a novel moment. The moment has to be novel for change to occur. The absence of novelty is sameness or identity. An entity cannot be self-identical from one moment to the next. This would imply an absence of change, thus an absence of time. The novelty of the entity, its temporality and change from one moment to the next, are codependent phenomena. The becoming creates the novelty as well as the duration through which the entity momentarily exists. Each novel moment is a constituent of an imaginative series over which the entity endures.

The genetic concepts can be related to those of change and time. Instead of the microgeny occurring over an objective duration that is linked to other genetic processes, one can say that the microgeny of the mental state does not fill time or 'take' time but is *time-creating*! The microgenesis of an object elaborates the time in which that object exists and is enjoyed. The momentary state is a universe of time and change. This world of the moment, which is all we know, becomes the setting for an illusory extension into longer or shorter time scales which then seem to occupy a portion of an external time in a relation of the phenomenal to the absolute.

There is a distinction here of a potential time that becoming creates, and a container time over which the becoming occurs. The distinction is important, as is the choice, because in making it one takes a stance on the frontier of the conceptual. Microgenesis is firmly committed to the subjectivity of temporal experience. Whitehead alluded to this distinction and proposed that mental space-time conforms to the dominant space-time of nature: he was led to the position 'that we are aware of a dominant space-time continuum and that reality consists of the sense-objects projected into that continuum' (ENP 102[3]).

Leaving aside the details of the theory and its clinical basis, which have been discussed at length elsewhere (LM), microgenesis can be characterized as a whole-to-part specification that recurs in rhythmic overlapping waves-fields, i.e., wave fronts, oscillators. The sequence is obligatory, recurrent and unidirectional. The cascade of whole/part shifts over evolutionary growth planes in the brain leads from a core in upper brainstem through limbic formations to the neocortical rim. The progression is from the intrapsychic to the extrapersonal, from image to object, from self to world. Consciousness is a configuration over phases in the same mental state, those that lay down, in succession, the self, personal space and the external world. This relation, as with every relation, is not instantaneous – in an instantaneity, relations are annihilated – but depends, as Whitehead pointed out, on a virtual duration that is derived from an imaginative reconstruction of the specious present (ENP 100).

Divisibility

In a microgeny the succession of phases is ordered from earlier to later (SP; TWMP),[4] though a complete traversal of all phases is necessary to establish a self, a world and a phenomenal now. Since the specious (phenomenal) present (see Figure 1.1) is extracted from a disparity across surface and depth phases, the disparity – in order for there to be one – obligates a realization of the entire sequence. But the phases do not have an independent existence until the becoming terminates, i.e., achieves an actuality, after which the phases constituting that becoming can be delineated. Without an actuality, the phases are 'out of time' and therefore nonexistent.

The brain state is indivisible, yet it is a complex entity. The indivisibility reflects the non-temporality of the succession of phases in each occasion. The epoch over which the phases are distributed, i.e., the subjective duration the entity elaborates, not its enactment in physical time, *does not exist* prior to its completion. The reconstruction of a phase-sequence is a retrospective act. The becoming is atomic, an indivisible unit of time. This temporal unit must first be created before its phases can be hypostatized.

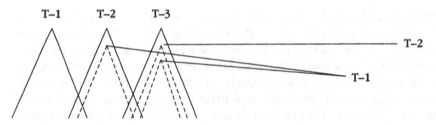

Figure 1.1. The state at T-1 is incompletely revived at T-2, less so at T-3. The duration of the present is extracted from the disparity between the 'surface' at T-2 or T-3 and the embedded 'floor' of T-1 or T-2.

Similarly, for Whitehead, the temporal passage in a becoming was not 'to be construed in the sense of a uniquely serial advance' (Whitehead, 1929 (hereafter cited as PR), p. 35). A becoming is not divisible into parts, though a gradation of phases can be described. The analysis of an entity is an intellectual act. Indivisibility is not a sign of simplicity. Indivisible objects are not basic entities. Indeed, there are no basic entities.

For some Whitehead scholars, the analysis of concrescence into phases, and the account of a sequence of prehensions, are inconsistent with the concept of a nontemporal becoming.[5] Sequence and phase are temporal concepts. But a phase does not count for something until there is an entity. An entity creates its phases no less than it is created by them. The completion of the becoming does not require that a set of phases complete its cycle but rather that the entity become itself, i.e., whatever it is. Waiting until the concrescence is complete before its analysis can take place is not a waiting for the details of the genetic sequence to be revealed. Successive states are 'called up' and ordered. The calling up creates the order. The ordering creates the temporal unit over which the calling up occurs, but not before a whole entity is achieved.

Clearly, a comparable paradox bedevils microgenetic thinking. The identification of segments in a continuum introduces an arbitrary demarcation. The continuum must objectify before the segments can be identified, but even then their demarcation is not possible. In the succession of phases, the direction is anisotropic. The formative sequence of brain evolution guarantees the direction of the actualization. Evolution and growth are constraints on the direction of process. If cognition is unidirectional, like phyloontogeny, the direction from past to present entails that a comparable sequence of phases in the microgeny should be discernible.

In each microgeny, phases are conceptual anchors in the continuous flux of change. The change within a microgeny is novel and indeterminate prior to an actuality. Every phase is a potential for the ensuing phase. A phase in transition is insubstantial, unbounded, like a wave in the ocean. The concept of phases or segments in a continuum, i.e., when boundaries are assigned, if taken too literally, may be irreconcilable with a whole-to-part process. Segmentation implies a concatenation or, if a continuity, one that is chopped into sections. Segments have delimitations yet can be overlapping; phases are less discrete.

Category and phase

The brain is an organic process through which actual or existent objects (acts, utterances, feelings, etc.) are created. Process is a pattern of change. What is unique about brain process is that it stabilizes change, or 'chunks' it into categories. The forming of categories achieves a stability of a natural kind that is unlike the artificial properties of intellectual analysis that have to carry the full weight of logical stability for objects that are otherwise unrepeatable. Categories are wholes to their members, which become wholes to subsidiary members, and so on, in a progression that is bottomless. The forming of wholes, or categories, is what the brain does best, and the effect is powerful. A real world and a constant self depend on it. That is why the illusion of stability is so pervasive and the dynamic of change so opaque.

More precisely, from the standpoint of conceptual processes, the continuum is a transition from a category (whole) to an instance (part) where the latter is the basis of another transition. The transition has the character of an emergence of whole-like parts from part-like wholes, where the wholes are not mere collections, and the parts are not definite elements but the potential to form subsequent wholes. The whole-part relation is a successive nesting that finally terminates in a concrete part, an actuality, that does not serve as a whole for a further transformation. An actuality is a concrete fact. An actual object is the finality for every succession of phases. The final transform completes the entity and thereby makes it real. The relation of whole to part is that of a recursive embedding of potentialities. One might imagine the pattern of concentric waves when a stone is tossed in a pool, but in reverse. It is questionable whether this relation is captured by the notion of phases or segments in a longitudinal series.

Past and present

The past is re-*presented* in the present. The development is wholly in the present but can be described as proceeding from the past to the present, loosely, from memory to perception. This progression is the reverse of the presumed flow of mental process in the research paradigms of experimental cognition and in neuroscience, where perception is held to precede memory, i.e., objects register and are secondarily identified through a match to items in memory. The immediate object is relayed to a past copy of that object for recognition. On this view, perception is inevitably passive. We never know what we are looking at, or what we want to look at, until after we see it.

Microgenetic theory entails that objects are recognized before they are *consciously* perceived, that objects are remembered into perception. A memory is an incompletely developed perception, while an object is a memory that has objectified or an image that has exteriorized. The sequence from past to present or from memory to perception corresponds with the direction of growth trends in forebrain evolution. For example, ancestral limbic systems mediate 'long-term' memory, i.e., meaning and experiential relations, whereas later evolved neocortical zones mediate the discriminant perception of external objects, i.e., the analysis of mental objects into (external) space.

The succession of phases in microgenesis is not to be construed as a conveyance or transfer of mental content from one stage to the next. A phase transformed was only the potential for that transformation. The transformed phase is a potential for the phase to follow. The final actuality actualizes the entire sequence. Preliminary phases in the object are ingredient and constitutive. We know this because damage to a preliminary stage can result, say, in a well-formed object (word, etc.) deprived of its meaning or recognition. Or, the conscious perception of the object's form can be 'lost' with good apprehension of meaning, i.e., meaning is 'encoded' before objects are consciously perceived. Here, the past of a present object – its recognition, familiarity, etc. – is realized into a present cognition, though the actual object that embodies that recognition has not been adequately discriminated.

An object is a process of momentary actualization. Each traversal from depth to surface is a minimal or irreducible unit of cognition and elaborates a whole unit of subjective time. The full, formative diachronic process, its temporal 'thickness' or extension, is the object. As in process philosophy, the essence of the object is its microformation; the object's being, i.e., its existence or being present or realness, is its becoming. A veridical object is the final thrust of becoming as process carries the past into the present. Every object is an assertion of the configural history of the organism. One can say that a present object consists mostly of the personal past of the observer. This past, or its abstract residue, is imminent, covertly, in every occurrent state.

In microgenesis, objects are generated from phases of potential to forms that become actual in order to become real. A pure subjectivity is avoided by the assumption that the material world modulates the generative process. What actualizes is a negative image of the entire realm of multi-tiered nature. Still the objects of perception are concrete images in the mental space of an observer. This differs from process metaphysics, which incorporates the object in the act of becoming, holding that an object is *given* or imminent in an occasion. The problem of substance-quality categories is circumvented by this move, though in my opinion at the cost of some coherence in the theory.

Every actuality revives the past as it actualizes. For the sake of a momentary appearance, the actual reaches back to ancient neural structures at the horizon of subjectivity. The past is out of time, and depends for its existence on the reinstatement of present experience. In the formation of the present, the past exerts a configural influence of which one is ordinarily unaware. This is the implicit past, i.e., personal and world knowledge, that is brought to bear on every act and object. The past becomes explicit when it achieves an awareness as a memory. Becoming explicit is becoming a fact of experience, even if what is explicit is a mental fact, e.g., an image, an idea however fuzzy, a proposition, etc.

A memory image is a present image of a past event. Were it to fully objectify, were the incipient present to vivify a recollection into a solid object, the result would be a present image of a present event. Whitehead wrote, 'there is no essential reason why memory should not be raised to the vividness of the present fact' (Whitehead, 1920), but to revive the actual *present* of a past state, i.e., not a memory of the past but a preceding actuality, would be to hallucinate with an object-like clarity. But of course there is nothing to say this is not the case with ordinary perception.

If an actuality is not achieved, the past remains forever past; it is excluded from subjectivity, thus from existence. Only in an actuality does the past become alive, and then not as the fact it once was but as an implicit constraint on a novel content. The continuum from implicitness to fact as a past (memory) grows into a present (perception), retraces the process of percept formation. The microgenesis of an object is a microcosm of its birth, life and death, a surge of the object into actuality out of abstract, timeless potential.

Change

Whitehead's metaphysics is a meditation of exceptional depth on the 'locus' of change in the mind and the world. An epoch of change is an epoch of time that is bounded by a past *out of time* that grows through subjective time into its forward limit with every epoch creating a present that is absorbed into timelessness for the next cycle of actualization. Time and change are a flutter of the imagination in the embrace of two eternities. This is so for process metaphysics where becoming is bounded on either end by an eternal (timeless) object. For microgenesis, the onset and terminus of the mental state are changeless boundaries encircling a process of change. The microgeny is a moment of time suspended between the limits of timelessness, like the experience of living, which is the dream of life that hovers on the eternity of sleep.

Every entity has a finite period – for a mind, a microgeny – over which it becomes. The entity then perishes and is replaced, as in the blink of the Brahma, by an oncoming epoch that is a near replication. The entity changes in becoming actual. The change is intrinsic to the actualization.

The actuality is epochal. The final object cannot be detached from all the phases in its becoming; indeed, it *is* those phases, so that change within the becoming is not apparent to a self that is deposited by the becoming, a self that is conscious only of a succession of objects, the *apparent* or illusory change from one actual object to another. The self is more closely replicated than its objects, which differ (come and go, change position, etc.) across the series of replications.

A replication is driven by the intrinsic constraints of the resting state at each phase as well as by the extrinsic constraints of occurrent sensation. A given state actualizes over the residue of a prior one; its thresholds limit the freedom in each traversal. The replication is never exact. The degree to which an actuality departs from a prior entity is determined by a number of factors, the flux of occurrent sensation, the baseline activity at each phase as it is activated, the decay of prior states and the emergent novelty inherent in the becoming process.

In microgenesis, change is *perceived* as a comparison across the successive occasions of an actual object. Objects are perceived as solid entities that change, not changes that assume the appearances of entities. The perception of change is a perception of difference not a perception of change. Genuine change is intrinsic to a given object; apparent (extrinsic) change is the perception of difference or a comparison across two changes. Genuine change occurs in the process of actualization through which a percept develops. The process leading to the object is the change from the object of a moment ago. Once the object actualizes, it no longer changes. The neural activity corresponding to the object is 'erased' in the brain as a path is prepared for the next actuality.

We perceive an entity as a solid because we need to perceive it that way in order to perceive it at all, and in order to survive. The stability of self and world has been achieved through a long evolutionary struggle. That is why we are here. The brain neutralizes change by transferring it from the time within objects to the space between them, displacing the change that is ingredient in the object to a surface interaction as another property of space. Genuine or non-illusory change is imperceptible for the reason that the change that is occurring can not be apprehended from a stationary viewpoint. The viewpoint is what the change is laying down.

Change is cyclical and pulsatile, a rising into actuality and a falling into abstract endurance. Wallack wrote that actuality 'jumps from occasion to occasion' (ENP). The jump from one object to another gives *apparent* change in consciousness. Genuine change and subjective time are generated by the actualization of a single object. Every change is a changed world. Change is not in the 'interval' between two actualities. The interstices of a series of microgenetic epochs are timeless, thus non-existent. The continuity across changes – the 'glue' of passage – arises from the timeless (changeless) 'gap' between actualities.

The observer has a perception of change across successive entities (worlds). Some objects seem to change rapidly, others not at all. Yet a butterfly on the wing and a stationary rock in the garden are each, as Whitehead would say, a mass of raging particles. The persistence of the rock, or its apparent lack of change, is not the absence of change but the relative similarity of its recurrence. The object keeps replacing itself and changes little in each replacement. With a butterfly, each replacement is a changed object. Change deposits the replacement in a series of novel objects. The perception of change, and the rate at which an object seems to be changing, depend on the resemblance of actualities across recurrences. With labile change across instances in a series of microgenies, the resultant entities appear to change quickly. Even with the minimal change of a rock, a recurrence is never exact. The light, the shadow, the perspective, the world around the rock, the world 'inside' the rock, everything changes in every change. The repetition of a becoming is always a new beginning. This is consistent with Whitehead's belief 'that *what becomes* involves *repetition* transformed into *novel immediacy*' (PR 137).

The meaning of change is linked to the meaning of time though change appears to be more fundamental because time is generated by change, i.e., an absence of change is an absence of time, and time more than change is mind-dependent. Change is what actually happens in the actualization of a given entity. Time is an emergent of a series of actualizations once the sequence of phases within an actualization has been established.

In process philosophy, there are two meanings of change. According to Leclerc, 'the kind of change involved in an act or process of becoming must be carefully distinguished from the kind of change constituted by a transition from one entity to another' (Leclerc, 1958, p. 79). Becoming is one form of change, transition across entities is another. The former is the process through which an entity exists, the latter is 'the difference between actual occasions comprised in some determinate event' (PR 73).

These forms of change differ in causal relations. Whitehead wrote that 'efficient causation expresses the transition from actual entity to actual entity; and final causation expresses the internal process whereby the actual entity becomes itself' (PR 150). The efficient causation (*causa quod*) of process theory, the existing state of affairs, corresponds with the apparent or illusory change of microgenetic theory, while the final causation (*causa ut*) of process theory, the state to be produced, the intention, corresponds with the real or intrinsic change of microgenesis.

There is no change of an entity in a becoming, for the becoming is the entity. The process through which the entity becomes is its change. Change within an entity is the process constituting that entity, so it cannot fairly be said that the entity changes through a becoming. The entity is not what it is until the becoming is complete. Once it is complete, it does not change, it perishes. Change and time in our experience of the world are perceived as related to the differences between objects. A difference involves a comparison. If the comparison is between simultaneities, the act of comparison introduces a succession. With time and change, the comparison is across successivities, i.e., from a prior to a subsequent state. The problem is exceedingly subtle. If the comparison is between entities that actualize but do not change, nor change once they actualize, i.e., if change deposits entities which themselves do not change, where is change if not in the observer, and what is an observer if not an emergent of change?

Time

Whitehead states that in scientific thinking, 'change is essentially the importation of the past and of the future into the immediate fact embodied in the durationless present instant' (Whitehead, 1919).[6] In process thought, however, the future is not imported into the present but is grounded in the present as a subjective aim. Becoming is asymmetric. The assumption of an anisotropy of time, along with the 'momentariness' of change in spite of the epochal nature of moments, aligns the theory with microgenetic concepts.

In process philosophy, the not-being of an object that perishes is not a nothing. It is absorbed into the permanent structure of the non-temporal world to endure as an abstract foundation for ensuing change. Actuality passes into timelessness; once timeless, the entity is eternal. In contrast, in the genetic process, temporal order is created out of non-temporality and devolves back into timelessness. The mental state spans a non-temporal inception that merges with a non-temporal outcome. Every mental state creates its own duration. Time inhabits many worlds, timelessness only one. Perishing and becoming are the ingress and egress of temporality as it pulses in and out of the same non-temporal ground.

In process philosophy, time is the conformity of successive states to their antecedents. The epochal nature of becoming displaces time from within an epoch to the succession of epochal states. Whitehead commented that the synthesis or realization of objects introduces temporal process. For some interpreters, such as John Cobb, time is in the transition; for others, time is in both the transition across occasions *and* in the becoming of the occasion itself. In microgenetic theory, time is only within the becoming. Time in duration, thus the past, present and future, precedence, etc., is created by a single state within which antecedents are embedded. The succession is necessary for the revival as an implicit 'layering' within the occurrent state. Time is not elaborated by the succession, but by relational features of the revival of antecedents within a unique occasion. A microgenetic state is both epochal and time-creating.

The perishing of a state is a dying back or attrition from the surface of the next revival. The entire state perishes but earlier phases are revived more readily than final ones. Conceivably, the earlier phases fade before the later, in the same sense that antecedent moments in the orbit of an electron no longer exist by the time the orbit is complete. The 'by the time' is the problem, for there is no time until the sequence is concluded. This epoch is required for an entity to become itself. Every happening within an epoch is out of time until the epoch is whole. Only in retrospect can precedence be established, so an entity not yet in time cannot perish, while once 'in time' it has perished already.

In each microgeny, earlier phases generate the past, later ones the present. Studies in process metaphysics have tried to disambiguate the before/after series from the past/present succession. One can have a before and after, but there is no past until there is a present for the past to be past in relation to. The actualization of the present transforms the before and after to a sequence from past to present.

When we say that decay begins with the present and leads depthward to the past, it is equivalent to saying that in the occurrent state, i.e., in relation to a present, only the past of a prior state is revived. Still there is a paradox in the association of early and late with past and present when the entire state is the present state, including the configural effects on that present of any and all prior states. The death of the present is the death of the entire state, but the greater reproducibility of the proximal portion of the state, and the graded revival of the distant, then the recent, past, give the impression that a loss of the distal segment is a loss of the actual present. The surface activity of the state appears to be progressively attenuated to make way for the next actuality, while the past or depth of the state is renewed with greater conformance.

Decay and revival point to incompletion of process. To say a prior entity decays in an occurrent state is to say it is partially revived in the ensuing one. The degree to which the state unfolds, and the recession of prior states within the present, give the 'specious' or phenomenal present. This duration is extracted from the disparity between the forward edge of the actual object and the 'floor' of the decay, or the 'ceiling' of the revival, of a prior actuality. The state is revived to a point where the immediate past is a content in 'short-term memory' buried in the actuality of the present.

The duration of the present in microgenetic theory is comparable in some respects to that in process philosophy, though the duration at issue is that of the subjective present, not the duration over which an entity becomes itself. A duration is not a stretch of time spanned by a perception but a virtual compresence of successive events in a 'concrete slab of nature', in which all events are simultaneous, and successive simultaneities overlap. Whitehead thought there was no explanation for this phenomenon. But a genetic approach to the mental state, in which the feeling of duration arises as an implicit comparison between the surface of the actual present and the revived (immediate) past within it, can provide an account of this aspect of the mental life.

In a succession of occasions, the initial state is revived less and less in each subsequent microgeny, eventually to recede into long-term memory beneath conscious accessibility. Every state unfolds over personal memory from the distant to the recent past. The present fades as a new present appears but phases of memory within that present are uncovered as if in a backward descent. In this way, the recent past and the momentary present, the recession of the old and the recurrence of the new, form the boundaries of a virtual duration that is the theatre of conscious experience.

Individuation

In the becoming of an object there is a progression toward greater definiteness. Whitehead wrote, in process 'the creative idea works toward the definition and attainment of a determinate individuality' (PR 150). In the process that generates an object, diverse entities *become concrete* by a coalescence or synthesis into a unitary occasion. Concrescence is the coming together of parts to form organic wholes. Microgenesis entails a progressive specification of parts that individuate *out of unity*. The fundamental direction is the analysis of spatial wholes into temporal parts.

The striving toward definiteness is the goal of evolutionary process. Form is shaped into objects by the elimination of the unfit. Cognition is microevolution. Before they even take hold, those routes of potentiality that could be maladaptive are pruned by sensory and other constraints to make way for what becomes actual. The endogenous generates a potential that is parsed to an object that survives the pressures of adaptation, i.e., the constraints on its development, to a fit with the sensory environment. The concept of parcellation in neuronal growth, the individuation in maturation of specific acts and objects, the analysis of gestalts into features, the relation of surround to center, theories of frame to content or context to item development in language and cognition, attempt to describe a wave of whole-part shifts through a succession of constraints on emergent form as a process in which diverse elements resolve out of organic wholes.

Mind and world

One can agree with David Bohm, that scientific objects are not *fundamentally* different from what happens in immediate perception (Bohm, 1965, p. 228). Mind is governed by the same laws as the material world. It is the agency through which the world is perceived and understood. Accounts of the object world are theories of the mind, and ultimate accounts of the physical describe universal properties of mind. A machine theory of physical matter leads to a mechanical or computational account of mind. A causal account of scientific objects gives a causal account of mental ones. Such a theory has difficulty explaining *patterns* of behavior unless those patterns are reduced to the effects of lower level elements, e.g., genes, chemical reactions, modules.

The inner is primary because a subjectivity at the mercy of experience would consist of random impacts. The subjective is not a construction but creates the assemblage of facts and contents on which its supposed explanation rests. Subjectivity imposes order on experience. The mind is an organism in constant struggle. The organismic theory of mind is harmonious with the concept of the world as a *creature*. Mind is a living organism that pursues its own nature. One can be lead to a theory of scientific objects through the mind or to a theory of the mind through science. Each theory should have a set of axioms the other can share. The starting point doesn't matter unless one starts with the wrong theory.

Whitehead based his metaphysics on quantum features of the material world and gave us the grounding of a philosophical psychology. A beginning with psychology, however, can lead to insights on physical process not anticipated by science. There is a deep consolation in the fact that the laws of mind and nature are reciprocally discoverable, and that both manifest the activity of thought. This leads one to ask if the becoming of material entities is an attribute of their existence in the mind or if their becoming in the mind is an instance of material becoming in nature. Put differently, is microgenesis a theory of mental process or is the process that the theory describes an instance of world process exemplified in the human mind?

Chapter 2
The Concept of
Momentariness

There have been few attempts to relate Buddhist thought to current trends in brain psychology. This is not surprising since the dominant force in contemporary psychology, that of a modular cognitivism, considers mental contents to be logical solids that interact in a function space, an approach that is incompatible with a metaphysics of relation and change, and equally inhospitable to process philosophy. The temporality that is central to Buddhist metaphysics, and foundational to all phenomenal entities, has been largely eliminated from the reified objects of psychology, which severs the temporal relations between modules and reinserts them in the connectivity.

The aim of this chapter is to bring to the attention of Buddhist scholars another approach to the human mind, one that has developed out of the study of the symptoms of individuals with brain damage. This account, the microgenetic theory of cognition, had its beginnings with the disorders of language (aphasia) that result from damage to focal brain areas. The various forms of aphasia were interpreted as anticipatory phases in the neurocognitive actualization or becoming of an utterance. Gradually, it became clear that the brain model of language was applicable to the account of action and perception as hierarchic systems of momentary actualization. Indeed, such a range of clinical phenomena could be explained by this theory that microgenesis appeared to constitute a general model of brain and behavior.

According to the theory, the mind/brain state is a continuous sheet of process from self to world, rhythmically generated out of a subcortical core and distributed over phases to a neocortical rim. The basic operation is a cascade of whole-to-part or context-to-item transforms in which parts arise out of wholes through constraints on emergent form at successive phases. The progression is from the archaic to the recent in brain evolution, from the past to the present – loosely, from memory to perception – in a momentary cognition, from unity to multiplicity, and from the intrapsychic to the extrapersonal. Mental process is uni-directional, obligatory and recurrent. The complete sequence from depth to surface constitutes the mind/brain state. On this theory, reality is not the starting point but the goal of an act of knowledge.

In the course of a reflection on the metaphysics of microgenetic theory, so many areas of correspondence to early Buddhist thought were evident – the arising and perishing of phases, the recurrence of moments, the phenomenal quality of perceptions – that one might have supposed it to have been the starting point of the work in neurology. Indeed, for many years after lectures on theoretical psychology, I have often been asked – to my dismay, I admit – whether my work on microgenesis was inspired by Buddhist philosophy.

Though roughly aware of the idealist tendencies in Buddhism, the momentariness and perspectivalism, it is only now, after the theory has reached a degree of maturity of exposition, that I have been motivated to explore the topic in greater detail to determine if there are some principles in common. The concept of momentariness that motivates this article is but one example of a convergence. However, even certain of the contradictions in Buddhism can be aligned with microgenetic concepts, for example, the concept of a succession of phases in a moment that is itself non-temporal.

The Buddhist concept of the moment

The Mādhyamika concept of momentariness or dependent co-origination (*pratītyasamutpāda*) holds that every entity is a series of momentary states bound together by similarity. An entity is reproduced through a replication of its states. Each moment comprises a state of the entity, though a complete entity is the result of an imaginative reconstruction over a series of states. The sequence of the replications is linked together in the mind through the rapid succession of similar moments. This gives the continuity of experience and the appearance of persistence. Satkari Mookerjee writes that the arrow in its flight 'is not one but many arrows successively appearing in the horizon, which give rise to the illusion of a persistent entity owing to continuity of similar entities'.[1] The judgment of similarity and the illusion that similar replicates constitute a single entity are subjective features.

In early (Abidharmika) Buddhism, every entity was conceived as a discrete element, and was held to emerge and perish in entirety. A relational theory of causal induction was applied to elements in sequence.[2] In a transition across elements, the entity did not become another entity but was replaced by the next in the succession. This was compared by T.R.V. Murti – like Henri Bergson, from a different perspective – to a movie strip with full stops and replacements, the observer filling in the gaps. The price of this atomism of the moment, however, was an inability to account for change. The entity, in Mookerjee's words, is 'destitute of all continuity', without a past or a future. If the continuity of changeless entities arises through a subjective addition, i.e. as a 'fiction of the understanding', how does subjectivity contribute change to entities that are otherwise changeless?

Continuity and comparison were problems for an encapsulated theory of the moment. In later writings, the transition from one moment to the next was conceived as governed by the causal laws of pure succession without an underlying substratum on which the laws were operative. Some commentators, most notably Ratnakīrti, emphasized that a unity of the self was necessarily prior to the individual momentary entities, in order to give them coherence. Yet if the self is a sequence of moments just like the entities it perceives, how does the fusion of moments occur? In an *anātman* theory (no self or soul), the coherence and the duration of entities cannot be explained unless an antecedent entity is postulated through which the phenomenal experience occurs. The assumption in Yogācāra of a consciousness anterior to experience attempts to resolve this difficulty but introduces problems of its own.

Every account of the phenomenal world requires an experiential thickness. At the most basic level, to say a cognition is not a simple entity but emerges into being impregnated with the impression of the previous cognition might address the experience of continuity, or at least the lack of awareness of discontinuity, but if all there is in the momentary state is a subtle departure from the state of the preceding moment – a departure that is non-conscious, for consciousness would entail a comparison across two states, not just a changed present state – there is still no accounting of (the illusion of) duration and identity.

Duration

According to the most central concepts of Buddhism, a moment (*kṣaṇa*) is not a time-slice that is demarcated and infinitely divisible; the moment is not infinitely brief but has a certain thickness that is fundamental. This thickness differs from a duration which is conceived as a psychic addition. The moment is epochal and atomic without incrementation; it is a minimal unit of time that itself is non-temporal.

The thickness of a moment is conceived as a durationless point in a continuum that has no definite boundaries. Stcherbatsky compared a moment – 'a momentary flashing into the phenomenal world out of an unknown source' – to a point in spacetime. The duration of the moment is 'the smallest particle of time imaginable'.[3] For Murti, 'a thing is a point-instant, having neither a "before" nor an "after"'. Murti writes that the moment has no temporal span, and thus no duration.[4] The duration of the moment is bound up with a theory of momentary states of consciousness that are the phenomenal equivalents of atomic point-instants. Consciousness and the duration of conscious experience are thought-constructions of the contiguities and simultaneities of the momentary flashings.

I would agree with those commentators on Buddhist thought who hold that duration is added by the mind to the series of changing points, not secondarily through a recombination, but implicitly in the act of perception (see below).[5] It is not sufficient, however, to argue that duration is a contribution of the mind to entities that are durationless. Such entities depend on the cognitive laws that govern the process of 'thought-construction' and these laws are as yet unknown.

In microgenetic theory, as in the writings of Henri Bergson, the duration of an actuality is the primary datum, the points or instants are artifacts. Without a duration, consciousness is not possible. The duration extends to the barest points since even these are conceptual. On this view, one needs an explication of the process through which a duration is established and from which the points are extracted, not an account of duration as a sum or an aggregate of points.

Phases

The moment arises and perishes. The thickness of the moment must accomodate its kinetic phases, i.e. the initial phase of arising, and the terminal phase of perishing. In a slice there is no width for even such a characterization; indeed, in a slice there is no dynamic, no relation. The phases of a moment are successive; the arising and perishing do not occur simultaneously. In some accounts, a nascent phase of arising, and a cessant phase of perishing, enclose an intermediate phase of abiding.[6] This intermediate phase is not a stasis but a segment of change that is neither an origination nor a termination.

In other accounts, there is no distinction of phases but rather a continuous perishing. Mookerjee describes a perishing that is intrinsic to the entity. The entity is its changing states. All entities come into existence and pass out of existence, not in the sense of a passing in and out of non-existence, though this is a topic of controversy, but of a passing from one state of existence to another state of existence. The change across different states does not necessarily entail a transition from existence to non-existence and back again. The perishing is the 'ground' of the arising. An entity cannot arise from nothing. An entity cannot perish to nothing.

The problem of arising (from nothing) and perishing (to nothing) has occupied many of the best minds in Buddhist thought, especially with regard to the causal process that binds one entity to the next. In my view, the problem of a perishing into non-existence is a confusion of an entity with an identity, the construal of a non-existence with a change in the identity of an entity, not the assumption by the entity of an altered state. The entity is no longer the same entity and in that sense no longer exists (as that entity), but one ought not to confuse non-existence with unreality or non-being. The non-existence of an occurrent entity at an antecedent or subsequent moment is its existence in another state, not its absolute non-existence. The misconstrual of non-existence with non-being results from a confusion of the momentariness of an entity with the persistence of that entity over the moments of its occurrence.

But within a moment, successive phases have been distinguished. The concept of a succession within a non-temporal span, or the concept of a non-temporal duration, is a paradoxical feature in the theory of universal flux. Unless every momentary state has some temporal width, a duration of some sort must be postulated to introduce process into instantaneous slices of physical immediacy.

In sum, an entity does not arise from nothing nor does it perish into nothing, but is a transition from one baseline state to the next. Something is retained in the causal transition which is the basis for the close replication. Otherwise, process would be chaotic. The antecedent state constrains the occurrent state which then constrains the subsequent state. Within each moment of arising and perishing, there is no nesting of nascent/cessant pairs. Nor can a static phase of abiding be tolerated. The moment is a dynamic cycle that constitute an indissoluble unit of existence.

Dependent co-origination

An event arises and perishes in relation to immediately antecedent and consequent events. Since every event depends on, or is conditioned by, these relations, and since every event itself is a relation, the event cannot be distinguished from the conditions that cause it. Thus, the event has no substantial or settled nature. One cannot say that there is the event! In Abidharmic Buddhism, this dependence was interpreted as a strict karmic determinism. The Sarvāstivādins were still bound to the Brahmanic idea of a self-nature (*svabhāva*) in which the essence of an element survives after its causal impact on the next event. The self-nature is the being or 'entity-hood' of the element. The Sautrāntikas denied self-nature.

The absence of substantiality or self-being in an element is a limit on its degree of actuality or realness. This creates the dilemma that either an element is substantial and real or impermanent and unreal. An event or moment as a segment in continuous change is impermanent, thus devoid of explicit content or essence, and in this sense, unreal or 'empty'. In Mādhyamika, these two perspectives are in dialectical tension. The middle path avoids the rigid determinism and changeless persistence (eternalism) of substantialist theories and the blind chance, continuous change and impersistence (annhilation) of relational theories.

While the development of the moment from a nascent to a cessant phase has often been described as a becoming, there is no process of specification from potential to actual comparable to that in Aristotelian thought or in process metaphysics. Rather, there is a progression from a phase of inception to one of termination. The concept of a potential is usually illustrated by an example such as the potential of a seed to become a sprout. This is not interpreted as the presentiment of the sprout in the seed, but in terms of a series of intervening causal pairs, i.e. from an earlier to a later stage in a series of states. There are occasional accounts of a potentiality that develops or 'ripens' into a cognition – by Candrakīrti, for example – but these do not seem to contain the concept of an individuation of entities from their generative ground.[7]

An essential ingredient in becoming is the many:one or whole:part relation. The frequent statements in Buddhist writings that there are no wholes, only parts, might be taken to indicate a rejection of the notion of becoming (of wholes into parts). Similarly, there is no progression to an actuality. The Buddhist moment does not progress toward realization. In becoming, in the process through which the individuality of an actual entity is achieved, there is a zeroing in on the definiteness of a realized particular. This 'movement' toward individuality is not present so far as I can tell in Buddhist thought. Indeed, the process of individuation is dispersed through the concept of co-dependence to a mutuality of inter-penetrated entities. Becoming is a process leading – in the metaphysics of Whitehead, for example – from a groundedness in the totality of eternal objects to the actuality of the one. In contrast, momentariness is the dependence of an entity throughout its transitional appearance on the totality of the universe at a given moment.

In dependent co-origination, entities are interrelated throughout their phases: every phase in the sequence from nascent to cessant is as actual or as illusory as any other phase. In becoming, the 'connectedness' or relation to all world entities of a given event – that is, the world-ground out of which an event is realized – is implicit only at the initial phase of potentiality. This phase is earlier in becoming and evolves to an actuality as an independent entity. Put differently, the dependent co-origination of a moment in Buddhism is a 'horizontal' segment of change in which all phases are equivalent, while the becoming of process theory is a 'vertical' unfolding out of time as the world is generated from potential.

These comments recall the process critique of the substantialism that is inherent in a 'vacuous actuality', where the entity fails to achieve a state of subjective immediacy, i.e. the perishing is the means by which an entity emerges from potential to become subjectively immediate. In the metaphysics of process thought and Buddhism, the actual does not attain the substantial in order to become real. The real is always relational in both the material and mental world. Indeed, in some schools of Buddhism, the relational is enlarged to incorporate the before and the after of every momentary event. This is the positive aspect of the relational. The absence of autonomy entails the interpenetration of every event by every other event. To paraphrase Whitehead, every actuality is somewhere, while every potentiality is everywhere.

One does not have to arrest the process of change to achieve an actuality. Mind and world are populated with seemingly concrete entities. These entities are illusory in their stability, but the presence of stabilities demands an explanation of how the flux is carved up into the manifold of appearances. It is not enough to assume that the stability of an entity is the result of a rapid succession, with duration and seriality achieved as in a movie strip by an illusory fusion of a rapid continuity of similar entities (frames) in consciousness, like a phi phenomenon. As noted, this will not work if consciousness is also momentary for then consciousness will require an explanation of its own capacity for fusion. The *anātman* position of Buddhism eliminates an absolute or enduring self that could account for the fusion of conscious moments. That is why the fundamental problem is the emergence of an entity. What is the minimal state of an entity – how can there be entities at all in a theory of continual flux, that is, how is an entity achieved in a moment that is durationless? Or, how are durationless moments compounded to the duration of an entity?

The phase of arising arises from a preceding phase of perishing. The perishing of the preceding moment is complete if the moment is atomic, but in some sense the autonomy of a past moment cannot be absolute without losing its continuity with the present moment. Disputation in Buddhist philosophy centers on the causal linkage between discrete moments, yet there is no theory of the moment itself. The moment is atomic, thus autonomous or indivisible, yet it is also relational, thus dependent and interpenetrated. How can the contradictory descriptions be resolved? The Buddha said of dependent arising or emptiness that it 'is the exhaustion of all philosophical views. I call incurable whoever holds emptiness as a philosophical view'. One is reminded of Wittgenstein's comparison of philosophy to an illness that is seeking a cure. But the relational theory of change on which emptiness is founded is a point of view. One gives up an attachment to other viewpoints to accept the view of emptiness, even if that view is viewless. Or perhaps my understanding is incomplete.[8]

The consequence of a momentariness that is pervasive and ruthlessly applied is that all knowledge is restricted to the present moment which, if sufficiently brief – the duration of a chronon, for example[9] – could not permit experiential awareness. We live, so to say, on a knife edge of the present, even if that edge is blunted, with each present conceived as an arising/perishing that is renewed as a near replica of the immediate past. The self is also a momentary entity that is conditioned by the antecedent state. Since each state exists briefly and then passes away, how can the self introspect on its own nature? This requires a distinction of self and object in a duration sufficient for an explicit contrast. The awareness of such a contrast involves a temporal relation that, in a momentary state, would be possible only if virtual or illusory.

Buddhism conceives a perception to be a constituent operation in the creation of the subject. The subject (self) cannot reflexively bind those perceptual moments that were ingredient in its formation. In early Buddhist thought, consciousness was held to vary with its contents. The awareness of a red object is a qualitatively different state of consciousness than the awareness of the same object when blue. In fact, the perception of a red object is a moment of a visual sensation, a moment of color, and a moment of pure consciousness arising more or less simultaneously. There is no consciousness *of* an object, no subject-object distinction, only an object-consciousness or an object perception and a state of consciousness in close association. What there is in awareness – awareness and its objects – is for the moment of that awareness all there is.

Time

A theory of time was not initially a significant part of the concept of momentariness. For the Sarvāstivādins, the elements were non-temporal but their function was temporal. K.V. Ramanan writes that, 'the unit of time is the unit of function . . . the minimum conceivable period (a moment) for the cycle of rising to function, carrying out the function and ceasing to function'.[10] Time is in the arising and perishing of moments. The Mādhyamika of Nāgārjuna did not attempt to explain the genesis of moments or provide an account of past, present and future in terms of a momentary Now.

Stcherbatsky notes that a point-instant 'cut loose from all imagination' is timeless, and durationless, but it can be viewed (intuited, apprehended by a mind) as a particle of time, the empirical origins of which are impossible to conceive. In the school of Dharmakīrti, the timeless point instant is a mathematical entity and the only reality in the universe. The point becomes a conscious present through the action of the imagination. The present, the now, fixes the past and the future and is essential for a direction of time. Stcherbatsky points out that the Sarvāstivādins raised objections to the convention that only the present exists, or that the past exists no more or the future is unreal for it does not yet exist.[11]

A major theme in the Buddhist theory of causation is the emphasis on temporal succession or cumulative causation, i.e. causal chains in the direction of 'time's arrow', with the penetration of an event (*dharma*) conceived as a penetration by the immediate past or the totality of present causes rather than a backward causation from the future. Vasubandhu said, except for one's own self, all the elements of the universe are the general cause of an event. Hartshorne has questioned whether in Buddhism there was a clear distinction of symmetric and asymmetric time. He wrote that in the tetralemmas of Nāgārjuna, the premises: earlier and later events may be mutually independent, interdependent, either or both, do not admit a time-asymmetric interpretation.[12]

In Yogācāra, because of the role of consciousness in the creation of the moment, the past and future are viewed as symmetrically embedded in the present state. There is a causal simultaneity. Past and future exist in the present by way of an impregnation or saturation of a perceptual consciousness-event. This concept is further developed in Hua-Yen, which entails a fully time-symmetric event causation, the past and the future collapsing on a single thought-instant in the present. The present is an island of immediacy surrounded and penetrated by an ocean of past and future time. This 'simultaneous mutual fusion' involves the emergence of a momentary part-event from the totality of transpired and to-be-transpired experience. This whole-part relation is one between a context or field and a core. This relation, which is central to microgenetic theory, has been compared to that between horizon and core in Heidegger or ground and figure in the Gestaltists.[13] The emergence of an occurrent particular out of a holistic surround of spatiotemporal totality, in which an antecedent and a subsequent temporality are compresent in the actuality of the present, is a type of whole-to-part specification not encountered in linear accounts of momentariness.

The occasional time-symmetric tendencies in Buddhist thought are not the analogue of the isotropic time of western science. Reversibility entails a linear time with a direction that is arbitrary. In those time-symmetric branches of Buddhist thought, time is conceived not so much as reversible as mutually determinate on the present. The causal inheritance from the past, the causal endowment to the future, are both simultaneously active in the formation of the present. The present is the resolution in actuality of an eternity of surrounding time.

In a becoming, a construct that is in a state of simultaneity with respect to its potential elements emerges to a linear order of those elements in consciousness. Events serialize as they become actual. Asymmetric time is generated out of timelessness, as wakefulness out of dream, with the terminal segments of becoming forecast in the antecedent phases of potential. Specifically, data present at the nascent phase of a becoming develop into the temporal facts of perceptual experience. This occurs through a transition from the wholeness of simultaneity to the partness of temporal sequence. This is not a symmetric collapse of the past and the future in the present but a creation of past and future in the achievement of a present entity.

The microgenetic theory of the moment

Microgenesis refers to the concept of the mental state as a dynamic traversal – becoming – over distributed segments or growth planes in the evolution of the forebrain. The traversal is a continuum of whole-to-part shifts leading from incipient phases of potential to a final phase of actuality. The actual is achieved as a cyclical derivation in which each

mental state, i.e. the full microgenetic sequence, displaces a preceding
state that is already in decay, as the developing state unfolds over the
residual activity. The state is like a fountain through which novel form
pours out from a subconscious core to a perceptual surface of world
objects; like a fountain, the arising of the present state issues out of the
waves of a preceding state that is receding in decay.

The process of becoming is conceived as a type of travelling wave that
sweeps from depth (arising) to surface (perishing) in an obligatory direction. The depth corresponds with the archaic in evolution and the experiential history of the organism in long-term memory, the surface with
systems of evolutionary recency and the immediacy of short-term memory
and perception. The sweep is from the past to the present with every
traversal depositing a novel occasion. The mental state is framed by an
arising and a perishing, but each phase in the state arises in the fading of
the same phase in the just prior state as it is revived in the next traversal.[14]

Mental states are atomic units that are replaced by near-replicates. Each
replicate changes slightly from the prior state in the course of its
becoming. In Buddhist thought, moments are continuous and recurrent.
Murti writes, 'change is the replacement of one entity by another; it is the
cessation of one and the emergence of another'.[15] In microgenesis there is
a fading of activity, and an arising on this fading of a new pattern of activity.
One enjoins, transforms and replaces the other. The phase perishes in the
sense it no longer plays its role in the prior state but it is not extinguished.
The perishing has a role to play; the phase survives, degraded, as a
baseline for the next traversal.

Change

The becoming of a mental state is a sequence of nascent/cessant phases,
both with respect to the wave-like pattern of the occurrent state, and the
background activity of the prior state. The perishing of the just-prior state
is the fading background on which the arising of the present state
develops. The neural activity at each phase in decay of the preceding state
is reactivated in the occurrent one. At each of these phases, a novel configuration develops out of the decay pattern. The synaptic relations among
the myriad neurons that constitute this pattern act as a template for the
ensuing configuration. In so doing, even as it decays, the pattern conforms
(constrains) this configuration to closely replicate the equivalent phase of
the just-prior state.

Change is the transition from arising to perishing and the replacement
of replicates, not the nexus between them. It is the configural deviation of
a replicate – a point-instant or a mental state – from the pattern of its
precedent a moment before. There is no awareness of change, only a
changed awareness; change lays down the awareness in a state that has
also changed. That is, the change of which we are aware, the flow of

objects around us, is not the change that generates the awareness, which depends on a comparison of states. This comparison is ingredient in the content of the changed state (of consciousness); it is not a contrast across successive states. The self and its objects in the present are not compared with those of a moment ago. The contrast is anchored in a duration that arises as a *virtual span across phases in a single state*. The mind does not suspend a past state for comparison with a present state. The past state is past, it no longer exists. Rather, past and present are re-created within the state of the moment with the events compared serving as boundaries *within* the momentary state, framing a duration that is extracted from a surface/depth disparity across phases.

The replacement of one state by the next is transposed by consciousness to an external connection of cause and effect. A substitution is apprehended as a linkage. Bergson wrote of this as a spatialization of time. A point that replaces itself is spatially extended as a two-dimensional line. The non-spatial point is visualized as a chainlike concatenation of states linked together by causal forces. From the standpoint of the mental state as an atomic whole, the linkage is an unreal void 'filled-in' by the changeless, timeless gaps between moments. Simple object causation is the filling in of the illusion of a linkage. The glue of passage is the absence of time (change) in these gaps. A fluidity of experience is achieved because the moments have no intervals. The intervals 'between the moments' do not exist because they are timeless. Each moment is an entire existence for the observer.

Arising and perishing

In Buddhist theory, a perishing gives way to an arising which gives way to a perishing. Does the transition from a perishing to the next arising differ from the transition from an arising to the next perishing? Is the cessant phase of one moment the nascent phase of the next? If so, from whence comes the autonomy of the moment? If change is a continuum of relations, how are moments articulated; i.e. on what basis is the flux of the world segmented into nascent/cessant points? No matter how sufficient the width of a given point-instant, its relations will still be transected to give the point, and the boundaries of the point will be arbitrary.

Is the difference between a nascent and a cessant phase that of the direction of process? That is, would an arising become a perishing if the direction of time was reversed? In microgenesis, there is a nesting of arising/perishing phases within the mental state, from the initial phase of arising, i.e. the onset of the mental state, to the terminal phase of perishing, i.e. the actual endpoint of the state. The arising arises from a baseline activity which itself arises from the oscillatory properties of neurons. The process goes from arising to perishing and is irreversible. There is an arising at the initial phase of a becoming, and an arising at each successive phase, but a perishing always provides a basis for another arising, i.e. a perishing is the basis for both the ensuing phase and the recurrent phase.

Buddhist theory appears unaltered by isotropic time. There is no reason why the process could not run backwards. In microgenetic theory, a perishing could not be an arising in a reversal of time. Change is in the direction of an arising to a perishing at each phase in the mental state, and from the onset of the state to its terminus. Process is organic and unidirectional. Nor is there a time-symmetry in process metaphysics, in which perishing achieves an objective immortality in the eternal ground out of which the next actuality develops.

Constraints and continuity

The wavelike process that is the vehicle by which the mental state is laid down is not a collection or composition of elements but a distribution into entities, a graded 'materialization' – a becoming real or actual – of temporal facts out of a nontemporal ground. The facts, the phenomenal objects of world and mind, are the distal specifications of earlier holistic phases. Within each mental state, the whole-to-part transformation is iterated - one should say, sustained, since it is a continuum - at successive phases through a cascade of transitions leading to the final actuality.

The mental state incorporates a sequence of phases from potential to actual. The phase of potential is an incipient object, in fact, many possible objects. The potential is a proclivity or a disposition, not a multiplicity. The tendency is from unity to multiplicity, not the reverse. The final object is the realization of that potential into diverse actual entities. What entities individuate depends on the intrinsic dynamic, but mostly on the sensory delimitation of the wave-like sequence of whole/part shifts. The brain does not provide a structure on which this process unfolds. The brain is a process through which acts and objects actualize. It is only phenomenally an object. This process deposits an actual object as a final fact of experience. The final phase can deposit the object world, or an interior world of dream, or an image that is embedded in the object world. The objects of perception, dreams, and the concepts and images of introspection, are all mental objects. These objects and the phases through which they are deposited constitute the mind/brain state.

The ground of the state, the resting activity of the basal phase, provides a context within which the state develops. Microgenetic states overlap. The replacement begins before the state actualizes. If the overlap were taken to the level of the point-instant, the corresponding concept in Buddhism might be the arising of a subsequent moment from the nascent phase of the preceding moment, i.e. before the perishing is completed, the next arising begins. The contribution of a T-1 phase in the decay of the prior mental state to the pattern at the same phase in the occurrent state at

T-2 is similar to the effect of the terminus of one nascent/cessant moment on the next arising. Through the overlap at each phase, a configuration replaces its antecedent and in turn provides a constraint on the state to follow. The influence acts at every phase in the generation of the state. A state is not activated by its antecedent. The antecedent does not provide the raw material for the consequent. Rather, the resting state establishes the boundaries of development and so limits the striving of process for the completion and fulfillment of those entities that are but dimly forecast in the phase of potential.

The state develops as a set of contrasts at successive phases. It is parsed by extrinsic sensory or physical 'input', and by the resting background of the antecedent phase. The becoming is propelled by the forward development and the pattern of intrinsic decay and sculpted at successive phases by the occurrent sensory field. Each phase in the occurrent state is driven by constraints, not direct causal effects. If a stone falls on the surface of a tranquil pool and causes a concentric ripple of waves, in what sense does the quiescent surface contribute, as a cause, to the pattern of the spread? If another stone falls, in what sense does the prior configuration cause the ensuing one? This problem is akin to that in microgeny.

A constraint is more like a condition than a cause. A constraint does not transfer something. A condition is necessary for an effect but is uncommitted as to the nature of this necessity. The concept of constraints as conditions that determine effects but do not go into them is consistent with certain schools of Buddhism. In this respect, the theory resonates with the distinction by Nāgārjuna, anticipating Hume, of conditions and causes, and the explanatory preference for the former because, as he writes: 'The essence of entities is not evident in the conditions'.[16]

The present state emerges out of its precursor. The foundational phases of the state evoke, in their development, ancestral phases associated with personal memory and world knowledge, and revive the past of the individual at every phase. This past constitutes the greater part of every cognition. Indeed, the inheritance of the past – the personal past, the life history – shapes the present at every moment. A perception is a memory redefined by sensation. Memory elaborates, sensation eliminates, all but what is relevant to the immediate surround. A microgeny specifies the world to a point where what actualizes is apprehended as external. An actuality is a memory, a dream, shaped by sensation to a space that is apprehended as real and extrapersonal. Conversely, a perception is a memory that escapes reminiscence by actualizing at a more distal phase in the microgeny.

Mere continuity or contiguity does not give a past. It does not give a duration or the recognition of an identity. A thermostat does not remember the past. The past exists for the thermostat, or for any reflex creature, as a changed present with no present other than the change. The imaginary present (ours, not the thermostats, which has no present) has appropriated the past, but this is a past that exists in the mind of an organism that is capable of generating a present, not the past (the before) of a system in which duration does not exist. A present is created by a mind. The point-instant cannot exist, as a present point, without an observer to be present for.

Buddhism postulates the moment, microgenesis the mental state, as the fundamental unit of experience. Each mental state is an encapsulated moment. In the autonomy of the state, the theory approaches the Buddhist concept of momentariness. The moment is a relation of phases that are internally continuous, but discontinuous with antecedent moments. Microgenesis gives an account of the origin of the moment as the basic unit of cognition. The process is similarly relational, the moment is equally discrete. There is an important difference, however. The summing of points to events is postulated in Buddhism to be an activity of the mind, while for microgenesis wholes are not constructed but partitioned. The moments of microgenesis are durations in which instants are artificial slices. The moments of Buddhist thought are irreducible points. Duration in microgenesis is computed from a disparity in phases; the phenomenal moment in Buddhism is a complex thought-construction of atomic point-instants. In Buddhism, points are constructed to phenomenal experience whereas in microgenesis they are the analytic termini of phenomenal wholes.

The mental state exhibits the same types of relations that govern the dependent co-origination of points. The moment is equivalent to the mental state, as the part is illustrative of the whole. The point-instant of Buddhist thought has also been related to the mental state. Mehta wrote, 'the steps of the paticca-samuppāda . . . apply in relation to each and every experience, and especially to each and every mental state'.[17] The mental state develops on the same basis as the point-instant. Relational dependence is universally applicable.

Like Indra's Net, holographic representation and fractal self-similarity, each point-instant is a microcosm of a phase in becoming, and each phase in becoming is a microcosm of the mental state. The points do not exist individually, nor do the phases, without a transition over the series. The state must actualize to become real, so its phases can also become real. As there is no arising without a perishing, there is no arising/perishing without another arising/perishing, so that any phase in the continuum depends on those phases in which it is embedded. The principle is the same in Buddhism and microgenesis.

Categorization and momentariness

Stcherbatsky wrote that the leading idea of Buddhism, the keystone of its ontology, is that there is no other ultimate reality than separate, instantaneous bits of existence, with all stabilities constructed by the imagination.

These thought-constructions of phenomenal experience obscure our vision of this underlying reality. For Nāgārjuna, the phenomenal world is as real or unreal as a point, and both are as real as any world can be. The relational dependence of events is true for any and every world. There is no substance, no foundational element; the points are recombined by secondary relations in the mind. But is not the very concept of the point-instant and its atomicity mind-dependent?

The mind-dependence develops on the concept of atomicity which requires a stability of process. Pure flux is not atomic. Atomicity is a break in flux. Buddhist theory is at pains to account for the autonomy of successive occasions. How is flux resolved with stability even at the level of the point? Is the relativity of the point compatible with its replication? If there is no explanation for how the flux is carved up, there is no account of novelty, identity, randomness, chaos.

The novelty of an occasion is in the relation of all points at a given moment that impact on that occasion. Is the 'sum' of all relations for each recurrence the sum of their causal bonds? The inter-penetration of points by other points – whether simultaneous or temporally displaced – is consistent with either rigid causation or continuous novelty. For some, the conditionality of dependence is a restriction on causal certainty. For others, the dependence is fully deterministic. The kinetic nature of the bonds within and between events implies that the emptiness of relationality could be the emptiness of contingency. The unconditioned is the ground of contingent entities. But if the relational is the contingent, its emptiness is not its contingency, it is its transitivity.

Change is in the relational quality of the points, stability in their capacity for replication. A constellation of moments that replicates itself with some consistency is perceived as an object. If replications are not exact facsimiles, they change. If they are exact facsimiles, they do not change. For microgenesis, change is the replacement by a *near* replicate. Consciousness and duration are created by the process of replacement; it is not consciousness that does the recombining. Indeed, in Buddhism, there is too much work for consciousness to do without some explanation of how the work is done. A theory of replacement is obligated to describe the change in the replicate; otherwise, the world is a static block. Perhaps for Buddhism it is, but then how and why does the illusion of change occur?

Process and stability are the problem of change and duration. The phenomenal present, the specious present of William James, is an illusion that develops in a single mental state.[18] The past event that is the hinter border of the duration no longer exists, nor do the intervening events, nor does the present event, the perception that is the forward edge of the duration. The external referent of a memory is gone, as is the referent of the perception by the time the observer perceives it. A perceived object is not 'on-line' with the perception. The perceptual event is a creation 'out of time' with respect to the external world, possibly with respect to the brain as well.

The world is the derivation of the conceptual into objects. Concepts are categories. The process of categorization is a primordial mental operation, a Kantian given, through which nature is articulated. The denial of categories in Mādhyamika – or the attempt to eliminate, through right action and meditation, the attachment to concepts or categories of thought – aims to strip off all thought-based illusions that obscure a knowledge of the real. Whether this is even possible is questionable, but the effort is justified by the categorical nature of all knowledge. It is likely that many of the categories of human knowledge approximate those of nature or tease her apart at her joints – otherwise we would not be here – but the categories are still mental. A moment or point-instant is a category, an isolate in the mind that takes on a contrast as a momentary focus. The emptiness of *pratītyasamutpāda* might be the pure flux that remains when the *concept* even of the point-instant dissolves. But then, the point-instant itself would disappear.

The stability of objects depends on a duration, which in turn hinges on the process of categorization. The process of category production is linked to the whole-part relation. The derivation of parts from wholes and the graded succession of whole-part shifts establish the pattern of mental process. The parts of a whole impend in the whole as the members of a category impend in the category. The whole is not a settled capacity but a potential for an arising of its parts. The parts are not actual items or constituents but virtual entities. There are no conclusive parts or wholes. A part that is generated can become a whole and perish for the sake of another partition. The final whole-like parts are what process deposits as the actual world.

Wholes and parts are relations. The relation of whole to part resembles the shifting contrasts and hierarchies of categories and members. This relation is fundamental to the experience of all mental and perceptual content. The shift from whole to part is the vehicle of change. The recrudescence of antecedent wholes that distribute into parts recaptures the context (concepts) from which the parts (objects) individuate. The view of mental process as the fractionation of wholes into parts – the final part-like wholes being the entities that process achieves, with every part emerging from an antecedent whole – corresponds with a basic theme in Indian philosophy, that unity underlies diversity, for example, in the concept of causation as an emergence of event-particulars out of a causal whole, or in the concept of a universal self or absolute nature that underlies the transient self and the phenomenal world. Mehta has written, 'from the many to the One; from the diversity so obvious to the senses, to the unity which is the fruit of inward realization: such is the general trend in religious thought'.[19] In microgenesis, this process is reversed. The specification of parts out of wholes is conceived as uni-directional, anisotropic, like the experience of time which is a subjective feeling that supervenes on the whole-part relation.

To say that a moment is a category implies that a moment in consciousness is a chunking by the mind of the stream of change. In fact, the appearance of a stream is elaborated by the moments. There are no present moments in nonconscious nature, yet there is momentary change. There are entities, there is change and, presumably, there is becoming, but there is no actual present. A moment in nature is not a now. In transforming moments to nows, mind exemplifies the becoming of nature. The change in a mental entity from its antecedents, the dissipation of an entity and its replacement by another novelty, the arrow of time that flies over the points as they are vanishing, the birth, growth and death of each moment, is the cognitive equivalent of the change that underlies passage in the material world. The momentary cycle of a point-instant, its arising and perishing, do not require consciousness for the expansion of that momentariness to a present moment. The present develops in a complex of nascent/cessant cycles within which the point-instant is conceptualized. All moments are not present moments independent of consciousness. If a point-instant were to be regarded as a present moment, one would have to postulate a consciousness ingredient in every entity no matter how particulate.

Chapter 3
Fundamentals of Process
Neuropsychology

This chapter explores some concepts and assumptions that underlie a metaphysics of process. The goal is to examine the formative dynamic of the mental state and extract some general principles or laws applicable to both cognitive and non-cognitive entities. In a process ontology, time and duration are central. The duration of an entity is its existence across phases of change. This duration is related to the conceptuality of the entity, to the past that is ingredient in every entity, to the connectedness within entities, to feeling, which is the unity of all actualities, and to the elaboration of a present moment, which is the starting point for a retreat from the complex to the simple.

General principles

In process metaphysics, the inner dynamic or genesis of a momentary actuality is the surge to existence through one cycle of change (see Rescher, 1992). Process theory proposes a single process of change common to all entities, not a bridge from the conceptual to the physical. The mental state is a model for this actualization (Brown, 1996). The laws of mentation are descriptions of regularities or patterns from which the laws of more fundamental systems can be derived. The pattern by which diversity is generated is a guide to what counts in the simplicity. The smallest particle should exhibit the pattern of the whole. The real is simplicity in existence, even if simplicity, as we know, is hard to pin down. The interaction of electrons involves clouds of virtual photons. Simplicity is no less elusive than complexity.

Despite arguments for a relational metaphysics, complexity is still regarded as a compounding of basic elements (Dipert, 1997). The materialist is correct in assuming that the laws of nature generalize to a theory of the mental state and that such laws are mind-independent, but not that these laws govern conceptual events or that conceptuality can be reduced to present day physics. David Bohm argued that neither particle physics nor quantum mechanics is a basis for a theory of mind. If physical objects are non-conceptual, we cannot hope to extend the laws of physics to those of cognition without introducing a novel principle at some point in the evolution of mind. If, instead, we begin with an elucidation of the laws of cognition and take the conceptual as primary, we may deduce some novel principles of physical nature.

Process monism

Process monism is the hypothesis that mind and matter, consciousness and microphysics, actualize generic process. The hypothesis asserts that there is one process in the world of which the physical and the mental are realizations; more succinctly, that the nature of mental process and the process of physical nature refer to the same process. This variety of monism differs from identity theory, which is just lower case materialism, in that the foundational process is no less conceptual than physical. The laws of cognition are amplifications in intricacy of the laws of nature. Process monism is a metaphysics of the antecedents of content or the laws of change that deliver a multiplicity of forms. The forms are concrete and phenomenal, but equally real and existing, the sum of what there is at the moment of their actualization.

There are similarities to the *concrescence* of process metaphysics. Whitehead wrote that a static monism of changeless forms was a reification of substance, thus incoherent, and argued against the concept on the grounds that actualities were concrete entities, not modes or appearances. However, the account of the mental and the physical as polarities in a single creative activity is close to a description of a monism based in *process*. Lewis Ford has written of Whitehead's *neutral monism*.

Concrescence refers to the concept of an actuality as a *growing together* of antecedent data into a novel unity. The antecedent data are prior actualities that have completed their becoming (Leclerc, 1958). In microgenesis, an actuality is a specification or analysis of a potential into concrete elements. The process is a phased individuation. The unity of the elements is their shared origin and history. The antecedent data of a concrescence correspond to phases in the brain state in microgenesis. The ground of a novel actualization is the fading configuration of the antecedent brain state out of which the ensuing state develops. The fading of the prior state is its incomplete reawakening, and the incompleteness of revival is its perishing for the sake of a novel occasion.

A conceptual world

Process theories from Buddhism to Bergsonism attempt to free the mind from conceptual bondage by a direct unmediated apprehension or intuition of time, duration, change, or the relationality of the absolute. But even if one could bypass mentation, the conceptual is present at the atomic level independent of cognition. The capacity of the mind to chunk items into categories that make wholes out of event particulars is the same capacity that gives rise to particulars out of categorical wholes. This capacity is at work in framing instants in durations and shifting successions to momentary simultaneities. The category is a virtual compresence of the temporal unfolding of its members that provides stability to the dynamic of incessant flux.

What is duration in physical nature? A first step is to entertain a concept of the physical world in which time is ingredient in structure. Time is relatedness; non-existence is an absence of internal relations. The trick of psychology has been to eliminate the relational for the sake of simplicity and reintroduce it later, as an external linkage, when the topic is in need of a more cogent understanding. Unfortunately, once an object is divested of the relationality that is the key to its subjectivity, it is hard to get it back inside. The relational has to be part of the original description.

Next, we must consider change or transition *within* the object as part of the definition of what the object is. An object, whether a mental state or an electron, is a transition gathered up as a stability. The duration of change – the set of transitions – comprising an object is not obtained by adding up the instants it incorporates. The instant is itself a duration that is divisible, and the duration exists only on completion of the change. Nothing exists in the figurative blink of an eye. The inception of the object serves as a past anchor to the fringe of an oncoming future. The actuality of the mental state constitutes one cycle of existence of the object. This transition to actuality is the relational process over which an object becomes what it is. The duration creates a stable object, while the stability prevents the object from dissolving into the emptiness of pure relationality.

Change introduces process, duration stabilizes change. The thesis that an object does not exist at an instant but requires a duration to become the object it is, that no object, even an electron, exists without a minimal duration, was argued by Whitehead, who wrote, 'in the physical world each epochal occasion is a definite limited physical *event*, limited both as to space and time, but with *time-duration* as well as with its full spatial dimensions' (Whitehead, 1926). The spatial extension of an object is its function as a container of temporal data. The relation of a cycle to its phases, or a duration to its instants, is comparable to the relation of a category to its members. If non-cognitive entities require a minimal duration to exist, and if duration is primitive categorization, conceptuality runs through all entities regardless of their ontological status. The process account of change entails that the durational existence of an object encloses one cycle of becoming that is replaced by another durational existence, and that this replacement is the vehicle of change in mind and world. If a non-cognitive object is a category of process independent of mind, duration could provide a nucleus out of which conceptuality might ultimately derive.

What does it mean for a non-cognitive entity to have a duration? Consider a spatial object. We think of objects as three-dimensional solids, and depth as a spatial attribute. Is the depth of an object in space not part of its progression in time? The oncoming face of a block universe frozen in time is a two-dimensional slice. An object with the thickness of a chronon (10^{-24} s) lacks depth (Whitrow, 1961). In order to introduce depth, a sequence of (fictitious) slices must accumulate over a series of point instants. The temporal dimension in depth is the succession of slices. Whitehead wrote, 'the extension of space is the ghost of transition'. Still, we can't resist spatial thinking. Concepts are spatial objects – the mind organizes experience into concepts and representations. Even the concept of an entity as the product of an atemporal becoming is a spatial or conceptual embrace of a dynamic set of phases rolled into a unitary object.

An atom is conceptual, not only in the imagination of a physicist but more deeply as a collection of phase-transitions over some minimal duration. Every existent exists as a categorical set of phases. The entity exhibits this structure when it becomes real. The entity becomes real independent of the observer. Space-time actualizes real entities. Perception fuses such entities into objects that persist. Percipience and categories are not imposed on the inner dynamic of the phases. The mind does not invent the conceptual, it discovers and elaborates it.

An entity is a phase-transition that spans an initial arising and a terminal perishing. The entity exists as a derivation or cycle. This duration is a kind of category. The phases have no existence apart from the set to which they belong. Since the succession is atemporal – the phases do not exist until the entity or category exists – their serial realization amounts to their concurrence in a momentary structure. In this and other respects, for example, in the way they spatialize process, an entity, a category and a duration are interdependent concepts. Entities are categories of atemporal parts. Concepts are categories that generate such parts, and the conceptual is, broadly, the conventional deployment of categories.

Memory and connectedness

We experience an integrated world. The perception of coherence is fundamental. Causation is our theory of this integration, but the intuition that the world is integrated is what gives the theory its force. The coherence is a *felt* experience of progression from an objective past to an actual present. We have the feeling of continuity over a certain thickness of change. This feeling is the connection of the particulars in a perspective to an ever-widening ground of potential. The connectedness is derived from the relations to antecedent phases imminent in the momentary life of every particular, coming into existence like the *anamnesis* of Platonic forms. Coherence is the thread of connectedness from the ground to the particular. We feel the coherence as an after-sensation of covert process when the diversity latent in a universe of being individuates. This individuation of the universe at any moment *is* its connectedness, and coherence is the feeling of connectedness projected as causal change in the world.

Central to the concept of mind is the connection of the present with its interdependents in the past, and the duration that arises in the context of this relation. Mind is not unique in this respect, but exploits what is fundamental in physical process. The succession of phases from before to after in a noncognitive entity is the seed of the past/present relation in consciousness. The phase sequence of the mental state, about 0.1 s, is a multiple of at least 10^{23rd} that of a chronon. Complexity enlarges the potential of an entity, thus the phases in its actualization and the duration over which they actualize. The further embedding of phases creates increasingly complex entities.

As the phase-transition expands and becomes sufficiently complex, the physical past becomes a cognitive memory. Antecedent phases that specify the past into objects are precursors of preliminary phases in cognition that import memory to the present. Early phases carry the distant past, later ones the present. The expansion of the complexity suffices for the enrichment of animal memory, instinct and habit, proceeding to human thought, meaning and remembrance. The momentary ancestry of an atom becomes the buried past of an organism, which gradually evolves to the implicit and experiential past, then, the recollected past of human cognition. The importation of antecedent data into the occurrent state empowers the present with an anterior dynamic and animates the causal history of the entity through the immanence of a past that would otherwise be lost in a linear model of change. Feeling runs through the connectedness as a river of unity that circulates through all occasions in nature.

From the complex to the simple

Conceptuality is the outcome of an evolutionary process that extends all the way down to the most basic entities. If we assume the presence of laws or regularities that govern all process, with complexity an elaboration of the initial-state pattern, not the result of novel principles added at later stages, we might expect to more readily discover the elementary in the complex than anticipate the complex from the elementary. And when we begin with the most complex state imaginable, that of human consciousness, the duration of the phenomenal present and the subjective immediacy of perception, we cannot help but arrive at a novel understanding of basic objects.

The most central aspect of a conscious state is the duration of a conscious moment. This duration is felt as a persistence over time, in contrast to the duration of a mental state which is imperceptible. We feel duration as a line in time, yet we are unaware that the perceptual contents within this duration are replicates that actualize over phases. The dynamic of individuation is obscured by its phenomenal products in the illusory stasis of the present. The duration of the present is felt but virtual, the duration of an entity, e.g. a mental state, is implicit yet real. The continuity of these durations – the present moment and the state that generates the present moment – is asymmetric since the former depends on the latter. The discovery that the duration of the present, the duration of the mental state, and the duration of non-cognitive entities, are on the same axis of evolutionary growth is a linchpin of conviction in the authenticity of process monism.

The fading (incomplete revival) of past events, and the disparity (comparison) of successive phases within the occurrent state, establish the boundaries of the present. Change is displaced from the becoming with which it is identified, to the world which it becomes. The becoming is not apprehended because there are no objects to be aware of until change is completed. Nor are there disparities in the phase-transition to serve as anchors for the distinctions of past and present, or subject and object, that are essential to consciousness. The becoming is absorbed into the present and obscured by the wholeness of the entity it creates.

In sum, evolution achieves complexity through an expansion of the succession of phases in the becoming of an entity. The entity is an individuation of the general to the specific. This process develops to an extent that the link of the actual to potential is obscured, yet even in excess, specificity accentuates the trend of diversity as the variation that specification brings to fundamental design. Mind is no exception. The incorporation of antecedent with consequent phases in non-cognitive entities in a relation of before and after comprises the duration of existence of an entity, or that period over which the entity becomes what it is. The phase-transitions that establish this duration expand in animal cognition to mediate the growth of more complex organisms. The antecedents of material actualities become the implicit memories of habit, recognition and learning. The embedding of phases continues in man to a comparison of successive phases in a single mental state. The before/after relation, which was previously implicit, becomes explicit in the comparison of a before and an after in the same act of cognition. The earlier (or the before) and the later (or the after) are the source of the posterior and anterior boundaries of the present. The transformation of the before/after to a past/ present/ future gives an evolutionary significance to the distinction of A and B series in time (McTaggert, 1927).

Feeling

Feeling is the secret unity of the world, the common heritage of every particular that gives coherence to the stream of individuals coursing to actuality. At every phase it keeps the world from flying apart. Feeling is the power in causation. In process theory, where entities are not the sources of causal power but the products of change, feeling is the necessity in the invention of novel facts. The world is a succession of becomings, and feeling is the power or pulsation of this process that distributes into a universe of entities as they are formed. Feeling is the complexity that shadows a particular when the diverse parts of an occasion are imagined in relation to the whole.

The affects and emotions of daily life are markers of mental process made perceptible by the objects it leaves behind. An object realizes a portion of a potential that is greater than what becomes actual. Similarly, an emotion is more capacious than the contents in which it is realized. Affects are shaped by the concepts they permeate, carved out as momentary contrasts in an ocean of process.

An object, as the substantive aspect of an occasion of fact, is the sum of its formative phases. Feeling, as the transitive aspect, is left behind when an object becomes real, a residue of the subjectivity from which a concrete entity resolves. Feeling is the direction, not the outcome, the conflict in the potential behind an entity and the process of its individuation, not the object of the resolution. We discover feeling in the surge to the present rather than the actuality in which it is ultimately embodied.

A metaphysics of feeling has its mundane expression in the emotions of daily life. Intense feelings anticipate the derived affects of consciousness. The evolutionary precursors of the will depend on a dynamic that foreshadows diversity. The reptilian brain, scarcely conscious in the ordinary sense, is capable of alarm and aggressiveness and a response to painful injury. A human newborn, conscious in the sense of an arousal or wakefulness, cries in fear or deprivation. Within weeks the infant expresses joy and displeasure.

Feeling is primordial. It begins with the forming of non-cognitive entities like a pressure to activity and realization, and passes by degrees to the organic. What has been termed the entelechy or élan vital is the first glimmerings of this activity in the earliest forms of life, where it begins and develops as the urge to preservation, which is the urge to replication or reactualization, to instinctual drive, the desires that are its aims, and finally, in man, to the strands of value that connect the object to the observer, at each phase binding the observer to his world as it binds the contents of mind and nature.

Mind and brain

For many philosophers, a theory of mind is anterior to science and largely independent of it, though the questions asked and the options considered are often driven by scientific findings. There is justification for this if by a science of mind one means the physiology of the brain and the firing patterns of nerve cells. The gulf between cognition and physiology cannot be bridged from either side. We have little precise knowledge of what neural or cognitive subsystems are implicated in the simplest behavior, not even a taxonomy that sorts cognition in a fluid and biologically meaningful way.

Historically, neurophysiology has been constructed on the model of the synapse. In a synaptic model, a nerve impulse at the axonal tip induces the release of transmitter which then ferries the impulse across a chemical junction to the dendrites of an adjacent cell where the next impulse is initiated. The output of one cell is connected to the input to another. The saltatory or boxcar flow of nerve energy is replicated in the input/output linkage of mental components and the transmission of discrete bits of information.

The synaptic model is most useful in the study of spinal motoneurons, where it originated to explain reflex activity, but the brain is not just a reflex organ of a more complex type. There is a fundamental shift in mechanism, one that requires a comparable shift in perspective, an enlargement of the physiology of discrete units to one of dynamic populations, perhaps closer to the discarded reticular or syncytial theory against which Cajal so capably argued (Cajal, 1954). Conceivably, the graded activity at the dendritic arbor might provide a more fruitful model of wave functions in a distributed system than the all-or-none discharge at the axon terminal (Pribram, 1971).

To an evolutionary perspective, the neuron is not a microchip in a computational network but an organism in a fight for survival. The life and death of a nerve cell is subsidiary to the fields or assemblies that form the environment of the cell, in which it matures, copes and perishes. The system is dynamic. Evolution is the story of such fields of organisms and their environments and the trends that determine the patterns of growth and function.

The link with evolution suggests that the patterns of physical change must anticipate the process of organic growth and decay, and that these trends continue into biology to become embedded in the process of cognition. The brain state is derived from growth patterns in simpler organisms which in turn are complex transforms of basic objects. Those trends common to growth irrespective of the complexity of the organism are the central lines of connectedness that unite change in all entities. One such trend that is fundamental in evolution and brain development, as in behavior, is the profligate growth of immature form thinned to parsimony in an accommodation to the constraints of nature. The mind is an organism that struggles to survive, while nature is an organism that shapes the mind to conformity. Some patterns of brain process in relation to growth include the following.

Abundance and economy

More individuals are generated than survive. There is a greater number and diversity than nature can support. The surfeit and diversity are parsed to conformity by their fitness to the environment. In evolution, the survival of relatively few adults is guaranteed by the abundance of many juveniles. This dynamic is replayed in the mental state, in the specification of a multiplicity in which the potential for the many is sacrificed for the survival of the one. The evolutionary struggle for fitness to the environment becomes the parsing of concepts to acts and objects through sensory adaptation.

What counts in evolution is the reproduction of an organism. An organism that is not replaced by a successor is an evolutionary casualty. The transition is from the organism of one generation to its progeny in the next. A self-similar replication is long-term survival; survival is replication in the short term. The transition across self-similar replicates each moment, and over the reproductive lifespan in the survival of the organism, is analogous to its replication from one generation to another.

In the course of its journey in life, the external environment is not the only world an organism encounters. The boundaries of inner and outer are indistinct at every level, from the habitat of the genes to an object in perception. The contribution of the endogenous is compromised by successive tiers of external constraints in the process of actualization. Survival is at stake at every stage from conception to replication in life and in cognition.

In mental process, what counts is the realization of a particular. The survival of a content in the mind, say an object in perception, is competitive success in the momentary genesis of form, success being the adaptation of the entity to the world through all the antecedent states over which it develops. Like the organism of which it is a part, an act of cognition is sculpted to precision by the elimination of maladaptive possibilities. The object survives this process and is able to self-replicate through another recurrence. The maturational process from the birth of the organism to the reproduction of another organism is accelerated in perception in the individuation of a concept in the mind to a fact in the world. Genes or organisms that are unfit usually fail to achieve rebirth. Mental contents that are maladaptive are generally eliminated before they are borne.

Potential and actual

The many-to-one process of evolution is recast in cognition as a becoming of wholes into parts. The becoming of an entity can be conceived as a many: one process when it is a transition from the *potential* for the many to the actuality of the one, and it can be conceived as a one:many process when it is a transition from the *unity* of potential to a diversity of realization. Potential is not a multiplicity from which a particular is selected but a unitary source of diverse phenomena. The account resembles the monistic theory of David Bohm, in which explicate entities actualize out of the potential of the implicate order (Bohm, 1980).

The transition from potential to actual lays down the self and mental content *prior* to the world of perception. This corresponds in process philosophy to the concrescence of a subjective aim in the prehension of a concrete actuality. An object is a public entity with an antecedent phase of subjectivity. If the subsequent phase of the object formation is also conceived as subjective, the conceptual extends into the world together with the subjective phase of all objective entities (Griffin, 1986). In every recurrence the mental state progresses from potentiality to determinateness. Creation mines the subjective for the novelty latent in potential prior to the definiteness of the actual. Creativity realizes organic form in a fluid motion over this continuum. Like the procreative, the creative revives the embryonic for a novel adventure in life.

Specificity by elimination

An act of cognition is a creative advance. Form is sculpted as it unfolds, not to uncover a configuration that is concealed, i.e. the configuration is not *there* to be liberated, but to chisel in relief the least contingent or most rigidly constrained line of development. Every phase in an entity, including the final phase, is conditioned to be what it is by what it cannot become rather than what it is determined to be. At successive phases in the mental state, sensory data nullify all routes of development save the one that occurs, the chosen path being the only possible path given the conditions that prevail. Ambient sensation combines with neural routines in habit, and synaptic biases in repetition, to configure and delimit the 'choices' latent in the opportunities for deviation each moment in the replication of mind and world.

The mental image of the world reflects the contours of reality that call it up. An object in perception is a momentary existent molded in space-time by physical sensation – itself a bundle of contrasts – to fit a permissible niche. We would not be aware of this were it not for dream, illusion, constancies and pathological disorders. But the adaptive nature of a perception does not imply that an object is illusory.

Since Hume, constraints and causes have been distinguished. There is power or necessity in causes, while constraints are passive inducers. Unlike causes, the content of a constraint is not carried over into its effect. In a contextual theory of change, in which entities are contrasts, relations are not arbitrary links but defining features. The relations define the entity, as an excitatory center is brought into relief by an inhibitory surround. The relations generate and demarcate a local density in the space-time continuum. This way of thinking introduces a universe into the formation of every entity.

Iteration

A basic object is a momentary existent that is reconfigured out of a universe of form by an incessant replication of its own ancestral line. The iteration is evident in the vibratory nature of the entity. Mental states are complex objects, with rhythmic or oscillatory properties. The states replace their predecessors in overlapping waves. The sequence of phases in the state is obligatory: an inception, a development, a perishing, giving way to the next state in the series. Not only the state but also every phase in the state is an arising and a perishing. The past is in the arising, the present in the perishing. The arising of the past configures the state to become a near facsimile of the prior one. A minimal deviation from an exact replication is the novelty in recurrence.

Growth is linear in appearance but deeply cyclical. We sense this in the reproduction of organisms, the periodic rhythms of nature and the body, the cyclical waves of cell migration that lay down the early brain, or the phyletic trends that reappear in maturation and the iteration of the mental state. With recurrence, history is repeated every moment with a slight inexactitude. In linear models of change, the past is revisited by an accident of contingencies. A linear system is continuously probabilistic. There is no definite pattern of change other than causation to keep the randomness in check.

Modularity and continuity

The dialectic of causal linkage and iterative growth is at the heart of the tension between a discrete theory and a continuum. This tension is resolved in the account of the mental state as a modular unit that is continuous *within* a becoming but discontinuous over replicates. An arising and perishing frame an actualization over indivisible phases. The indivisibility is essential to the wholeness. A division would require the atomization of potential into temporal instants before the temporality has been established. The mind/brain state is not the sum of a series of separate slices; the continuum cannot be sliced, and the phases are all necessary before the state exists.

An entity is an atomic or modular unit of change, but a complex entity is not a compilation of basic modules. Complex systems are expansions of basic entities and remain atomic though the duration of their actualization increases. Atomicity is not aggregated to composite entities but enlarged from within. The continuity of an atomic unit is by way of a covert replacement that is constrained to replication yet is sufficiently different from its immediate ancestor to have the appearance of a chaining through causation.

Derivation, not interaction

Every object has a momentary career that forecloses interaction with other objects at the same plane of completeness. Like parallel lines, mature objects perish but do not intersect. The one-way action of interaction is from depth to surface by individuation. The concept of interaction, as in billiard-ball causation, assumes a collision at the boundaries of object-appearances in the world or logical solids in the mind. Whether interaction posits an exchange of energy, or a transfer of information, or a fusion to a new entity, the entities are assumed not to change until they interact, other than in motion which is a change in relative position.

For process theory, the motion of one object in relation to another is a change in object relations that occurs over an intervening series of derivations. Process and change are from antecedent to consequent in the same mental state, not across two consequents at the same terminus. In its actualization, every entity constrains and is constrained by all other entities. What passes for interaction is the mutual constraint and connectedness within and across forming entities in the course of a momentary becoming.

Part II
Consciousness

Chapter 4
Psychoanalysis and Process Theory

All our scientific and philosophic ideals are altars to unknown Gods.

William James

The *Project* was Freud's final and self-admittedly unsuccessful attempt at a neuroscience of the mind before he abandoned neurology altogether for a clinical theory of mental life. The work was clearly a reaction against the logic of association psychology yet subtly incorporates to its detriment many of the basic assumptions it sought it supplant. The perceived failure of the *Project* was, of course, for the neurology, not for the psychoanalytic concepts, which continued to shape much of his future work.

This chapter takes up the central issues of the *Project*, the distinction of separate systems for memory, perception and consciousness, the postulation of unique cell types to subserve these different functions, and the attempt to imbue them with a makeshift dynamic through the concept of energy flow and redistribution. The dynamic was makeshift because a fully process-based approach would interpret these systems as segments or relations in the flux of change. It is proposed that microgenetic theory can provide such an approach, in which memory, perception and consciousness are conceived alone a continuum of change in the actualization or becoming of the mental state.

Freud's Project

The *Project* was an attempt to resolve a non-local brain anatomy with psychoanalytic concepts (Pribram and Gill, 1976). In this, it does not so much mark off a new direction in theory but is a transition from an associative anatomy, such as that of Carl Wernicke, whom Freud (1891/1953) criticized,[1] to a purely psychological association that continued the trend of turning fluid concepts into solid entities in complex interaction. The vocabulary was still that of Wernicke, even if the concepts were divorced from localization. For example, Freud argued that thoughts became conscious by associating with verbal images or achieved

47

expression by an association of sound and word images with motor speech images.[2]

Freud was also strongly influenced by the top-down hierarchy of Hughlings Jackson, a ghostly presence in the monograph, more conceptual than anatomic, in the accounts of inhibition, censorship, and especially repression. For the 'horizontal' connections of Wernicke and the 'vertical' inhibitions and re-representations of Jackson, Freud substituted a conceptual schema of causal interaction between affects and ideas independent of their anatomic substrates.

With regard to the concept of repression, that certain ideas remain unconscious is not a justification for a mechanism to repress them. It is implausible that a discrete idea can be isolated from its context and selectively decathected or screened from consciousness. The selectivity requires a micro-decision, and the mechanism – actually many mechanisms, energic flow, decathexis, anticathexis, substitutions, etc. – must explain how any content becomes conscious regardless of its affective charge, and what 'becoming conscious' entails. Nor is the concept linked to memory and normal forgetting. One lesson from neuropsychology (e.g. split-brain cases, blindsight, and everyday observations in the clinic) is that contents may not surface to awareness without the need to postulate an active process that prevents them from doing so.

We do not understand the 'laws' that govern the derivation of contents from an unconscious (Ucs) to a conscious (Cs) state but there is no need to postulate a block or an inhibitory force. The affective tonality of an Ucs configuration will play a part in determining whether that configuration will actualize. In some instances, the affective tone may facilitate the derivation, in other instances it may impede it. The likelihood of a potential content becoming actual (that is, whether or not the configuration will undergo an individuation to a Cs content) is a function of the degree to which the content is congruent with the occurrent state in the course of a development through semantic fields or graded networks of meaning relations.

The association of Wernicke and the inhibition of Hughlings Jackson derailed the development in psychoanalysis of a process-oriented neuropsychology,[3] a task taken up by others such as Paul Schilder and David Rapaport. Freud did, however, recognize *the* fundamental problem with association, that it postulates mental solids in interaction where the dynamic – the change in the system – is collapsed to a mere connection between two causal surfaces. The concept of energy or quantity (Qη) was thus invoked to infuse inert elements with an extrinsic dynamic that was lacking in association models. I say extrinsic because even if drive energy is construed as intrinsic or drive-originating,[4] it was still an additive factor to neurons that were basically passive receptacles. The energic theory was an *ad hoc* dynamic in a function space of mental solids an association model shored up by extrinsic energy, not a dynamic theory *tout court*.[5]

In search of process

The need for a more dynamic account is seen in Freud's distinction of the endogenous psi neurons for the storage or capacitance of memory and the exogenous phi neurons for the immediate discharge of perception. Time is the central issue. The distinction between memory and perception is that of persistence and instantaneity, that is, an enduring event in a continuous self and an actual event in a fleeting present. Persistence need not raise a temporal problem unless the persistent entity is viewed as changeless. Subjective time involves a duration over which the before and after of physical passage are suspended. A change in the setting of a thermostat might be construed as a trace that endures, but does not require a duration throughout which a trace is active. Time becomes a factor when persistence is a sustained presence *in awareness* of the persisting entity.[6]

Persistence and instantaneity are temporal concepts. A persistence in memory is a change over a duration. A duration is a relative length of time. An instant is a shorter length of time within a duration. Persistence and instantaneity are related as a duration is related to an instant. As with the postulation of $Q\eta$, they reflect the need for a temporal dimension, a need that underlies the search for a *process* theory, but the postulation of separate neuronal systems to subserve these functions, and the system consciousness (*Cs*), is a throwback to associationist thinking. The resolution of persistence and instantaneity is achieved by a process through which memories *become* percepts, or through which the duration of personal memory *becomes* the now of the present moment.[7]

To sum up, the major neuropsychological problems facing Freud were to resolve mnemic potential and persistence with perceptual actuality, to rescue a dynamic framework from the static architecture of association, and to introduce a temporal or processing dimension into a three-dimensional space of conceptual solids. The failure of the *Project* was a result of the pervasive influence of an association psychology conventional for its day – even to this day – when a psychoanalytic neuroscience required a completely new neurology.

Brain-state and psyche

The seed of this new model can be found in the concepts of a progression from the unconscious (*Ucs*) to the preconscious (*Pcs*) to *Cs* and the association of *Pcs* cognition with hallucinatory imagery. Freud wrote that 'remembering is a precondition of all testing by critical thought.' Thought and memory are closely related. Thought depends on new contents or recombinations, while memory depends on a feeling of familiarity. One could say that thought is productive memory, while memory is reproductive thought. Freud maintained that hallucinations are recollected perceptions; thoughts are substitutes for hallucinatory wishes and dreams are wish fulfillments.[8] Thoughts and dreams are partly acts of recollection. This is not true of perceptions, which discharge immediately, and deposit as memory images. 'The primary memory of a perception', Freud wrote, 'is always a hallucination'. On this account, first the perception registers, then it recurs in the form of a hallucination. This goes back to the neurological idea that perceptions are stored as images which can be secondarily revived, as in hallucination, or associated to new percepts for recognition.

In my view, the idea of perception as a peripheral or input event was the critical error that doomed the *Project* to incoherence. What Freud should have argued, but could not because of the limited neuroscientific knowledge of the day, was that not just thoughts but perceptions as well develop out of hallucinatory memories – that is, that memory and imagery are preparatory or pre-processing stages in perception, that perception is imagery exteriorized. Objects are not perceived in the periphery of the mind/brain to then enter a memory store for later matching (recognition) and revival (hallucination, thought). Rather, *we think or remember objects into actuality*.

The progression is from *Ucs* primary process/mnestic-like function with affective and experiential relations predominating, to more abstract categorical knowledge, to pre-attentive (*Pcs*) gestalt-like configurations that are analyzed (specified) into perceptual awareness. Sensation constrains this progression at successive moments to model an object in perception. That is, a fully endogenous process of image generation is sculpted by sensation to represent an exteriorized world (Figure 4.1). The process of becoming involves a qualitative change from one segment to the next rather than a retrieval or copy mechanism or a sequence of fractal-like self-similarities.

LEVELS IN THE MIND/BRAIN STATE

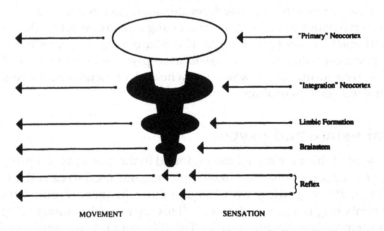

Figure 4.1. Microgenetic schema of the mind/brain state. A combined (simultaneous) act-object emerges in a brainstem reticular pool surrounded by sensory input and motor output. The derivation over endogenous phases from the core to the world surface is the mind/brain state. This 'psychic capsule' is surrounded by successive physical tiers of sensory constraints and motor keyboards that are extrinsic to cognition.

Freud did not identify primary process thinking with the *Ucs* for *Cs* could attach to primary processes, yet studies of pathological cases show a decreased *Cs* of content in relation to the degree of deviance in word or object meaning. The more deviant the meaning relations or the more primary-process-like the entity, the less likely the subject is to be aware of the deviant content. The lack of error awareness or *Cs* of content is in relation to the semantic deviance of the content. This implies that when *Ucs* contents rise into *Cs* they undergo qualitative change and are not mere copies of the same contents in the *Ucs* (see below).

In later writings. Freud wondered whether a transition from *Ucs* to *Cs* involved 'a fresh record – as it were, a second registration – of the idea in question (that is) a fresh psychical locality . . . (or) a change in the state of the idea . . . involving the same material and occurring in the same locality?' (Freud, 1953 (hereafter cited as SE); 14, p. 174). In either case, the idea itself is unchanged, it is a question of re-registration from *Ucs* to *Cs* (the re-representation of Jackson), or *Cs* attaching to an *Ucs* idea. Freud did not decide between these alternatives, noting the possibility that 'the difference between an unconscious and a conscious idea (may have) to be defined in quite another way' (SE 14, p. 176). The transposition of an idea from *Ucs* to *Cs*, or the cathexis of an *Ucs* idea by *Cs* is still too mechanistic, for the idea is treated as an entity that is a target for energy or an explicit content that is shuffled between levels or components.

In fact, an *Ucs* idea changes in a continuous actualization as the mind/brain state *develops* from depth to surface, from events that are memory-like to those that are perception-like, but with mnemic and perceptual content recognizable at each moment in the development. There is a depth and a surface to memory *and* perception. For example. long-term, short-term, and iconic memory are ways of characterizing a perception at successive phases in its realization. Similarly, the memory-like phenomena of dream, hallucination, and imagery are phases in a developing percept (see below). This fact is obscured by the belief that memories are stored subsequent to perceptual encounters. But every present object develops out of the past. Memory is brought to bear on perception, not after it is recorded, but in the original process through which the percept is formed.

Some main problems with association theory

That Freud could not transcend the theory of his time in spite of his disaffection with it is hardly surprising. Association thinking has had many detractors over the years but few alternatives, and in varied disguises it continues to dominate current-day discourse, for example, in the localization models of neurology or the componential models of psychology and philosophy of mind. It is therefore worth noting some of the problems with association logic and its permutations in Freud's work.

Regardless of whether the units of the theory are physical centers and pathways, or modules and networks, or physical elements such as representations, agencies, images or propositions, or 'processors' such as buffers and stores, the units are treated as separable, interactive domains of function in a conceptual space where change is the flow of information from one component to another. A component has an output or receives a charge or a cathexis. Since the change is located at the interface of an input or output, it is extrinsic to the component. The input fires the component, or is attached to it. The output is the change the component induces. In either case, however, the component itself does not change.

The flow from one element to the next gives a concatenation similar to a reflex chain in which events are simultaneities at an instant. The approach is homuncular because it requires decisional nodes at successive branches to direct or reroute the process. In psychoanalytic theory, there are homunculi at successive levels, whether id, ego, and superego, or the more elemental units that are the building blocks of these entities. In cognitivist accounts, the homunculi are, minimally, the box-like elements in flow diagrams as well as the integrators that bind the ensemble of elements together.

Mind is the sum or aggregate of a multitude of elements in a given slice of process. The unity of the mind is an adventitious feature, achieved either through recombination or through the coincidence of timing of distributed events. The coincidence is explained by the physiology (e.g. a binding mechanism) and is not an intrinsic property of the psychology. In my opinion, the hypothesis of a mechanism that binds a multiplicity of elements together is an artifact of the time elimination that was necessary for the isolation of the elements in the first place. Temporal relations are inserted in the interstices of the elements to account for their linkage when those very elements depended for their demarcation on the abolition of temporal relations.[9] In the search for such a device, the temporality disposed of early in the psychology has returned to haunt its theoretical maturity.

Componential 'theory' is not necessarily computational but computations are invoked to account for the manipulation of elements or symbols. Typically, the associations are learned while the computations are innate but this distinction is not absolute. Cognitivism fills in the black box of behaviorism with a mental reflexology in which the computation does the work that was previously the responsibility of the association. A computation imposes an extrinsic change in the form of a rule on an otherwise stable element. A rule induces a transformation but is distinct from the entity that is transformed.

For interactive theories, and psychoanalysis is such a theory, the foundational units of a mental state are atomic entities that recombine rather than holist entities that individuate. This entails a perspective that gives parts with boundaries that are artificial and wholes that are mere compilations, not the process through which parts *become* wholes, or the reverse. Componential theory is not inconsistent with the concept of *Ucs* mentation since mind is inferred even in the absence of *Cs*. The assumption that a computer can think entails that *Cs* is inessential to human or machine cognition or that it can be assigned to some components and not others.

Toward a psychoanalytic neuroscience

The concept of a perception as a terminus in the actualization of a mind/brain state leads to a complete rethinking of the nature and direction of mental process. The nature of the process is a microtemporal (microgenetic) becoming of action and perception from depth to surface through a series of whole-part or context-item shifts, while the direction is anisotropic and obligatory, involving a progression from the intra- to the extra-psychic, from self to world, from meaning to fact, from concepts that have not yet actualized to the 'reality' of world objects. In this process, potential at the depths of the *Ucs* transforms through *Pcs* and *Cs* segments to become a discrete event, an image, a percept, a momentary world.

The mind/brain state is replaced in a series of overlapping waves, each developing 'bottom-up' in a fraction of a second, from ancestral formations in the brainstem to neocortical systems of evolutionary recency. Freud's speculation that a periodic neuronal discharge could provide a temporal dynamic to the quantitative model anticipates the concept of a pacemaker in upper brainstem that initiates the mind/brain state.

Mental contents are deposited over successive layers of space representation. An object is a contrast in a world image that is the field of a perception. This field is one of several that are arranged in layers, e.g., one populated by dream images, another by concepts, yet another by objects. The object is not the final effect of a sequence of microtemporal phases that constitute prior causes, but 'contains' the full set of its formative layers. These contribute to the recognition, the familiarity and the meaning in the object, as well as to the spatiotemporal context in which the object appears.

Presumably, it is only the final object, not its earlier stages, that corresponds to a 'real' object in the world. There is no external world of object meaning to which the concept of the object refers. Of course, this is a matter of belief not fact. An object in perception – the chair before me – is a mental event that is inferred to mirror a real object, its presumed cause, not through the construction of an object by a coming together of sensory 'bits' but through a process of sculpting as a configuration individuates through the constraints of sensory excitation. The postulate of an object in the world that corresponds to our perception of it is a speculation on the sources of this excitation, just as the latter is a speculation on the immediate origins of the perception. We must keep in mind that a chair corresponds to a spatiotemporal pattern of activity in our brain. It is a subjective image that has exteriorized in the mind's 'external' space, a mental entity like an image or idea, but more definite, concrete and pictorial and situated in a different plane of mental space. When I scan a mental image or reach for a chair I am swimming in my own imagination.

In this view, self and object are inextricably wedded in the same field of brain (and mental) activity with Cs the relation of self to image, or object. in a continuous sheet of mentation. Cs is generated[10] in the course of object perception as a link from the self to the images or objects that are its distal phases.

The microgenetic approach to memory and perception

The distinction of memory and perception was the centerpiece of the *Project*. Freud maintained their separateness in cognition and in anatomical substrates, and this has continued to be the standard practice to this day, not surprising in view of the common sense account of memory as a bank from which items are retrieved and perception as a contact with the world. This account has only grown stronger as the fragmentation of mnemic and perceptual systems into sub-components serves to prop up deficiencies by positing additional components just where theory is found wanting. What is needed is an understanding of the *process* of remembering and perceiving.

Figure 4.2 depicts the microgeny of a visual percept, but it could as well depict the staging of 'components' in memory, since memory and perception are different aspects of a common process. A perception begins with a possibly two-dimensional spatial image in upper brainstem. This construct is transformed by limbic structures as the preliminary configuration of an extrapersonal object individuates out of a volumetric, viewer-centered space in association with long-term experiential or personal memory, symbolic and affective relations. The configuration is then transformed through a Euclidean space of object relations associated with short-term or working memory within the pertinent modality. The final analysis of the emerging configuration into the featural detail of a fully exteriorized, independent object is associated with the stage of iconic memory The transition is from the enduring and largely Ucs patterns of experiential memory, through a Cs phase of working or STM memory, to the largely non-Cs phase of iconic memory. The initial phases are Ucs because they are prior to the self and Cs. The terminal phase is non-Cs because it occurs in relation to an object that appears external to the mind. In other words, like an object that has exteriorized and is perceived in a separate physical space, that phase of the mind of which the object is but an element also exteriorizes and is apprehended as the public domain.

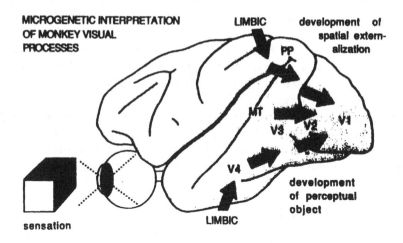

MICROGENETIC INTERPRETATION OF MONKEY VISUAL PROCESSES

LIMBIC   development of spatial extern- alization

PP

MT   V3   V2   V1

V4

development of perceptual object

sensation   LIMBIC

Figure 4.2. Microgenetic theory reverses the standard direction of processing in percept formation. The inception of the mind/brain state in upper brainstem is not shown. The diagram, taken from Deacon (1989), illustrates the progression from limbic formation to primary cortex. The flow from V4 to V1 is consistent with the laminar pattern of neocortical connectivity.

These phases are recaptured in the varieties of perceptual experience. The phase of personal memory reappears in dream and scenic hallucination, intermediate phases appear in association with thought and memory images, and distal phases with afterimages and eidetic phenomena. The different types of images are segments in the realization of the putatively 'consolidated trace' of long-term memory into the 'fleeting impressions' of perceptual objects.[11] The preliminary phases of LTM or dream are associated with largely *Ucs* or *Pcs* cognition. The intermediate segments of working memory, STM, memory or thought images and inner speech are associated with *Pcs* or *Cs* cognition, while the distal phases of iconic memory and after-imagery constitute the 'physical' rim of cognition as it realizes an exterior world. *Ucs, Pcs, Cs* and the perceived world are successive segments in a single mental state.

A perception that is revived is a memory or hallucination, but if it is revived 'too far' it again becomes a percept. Similarly, a perception that does not fully 'come up', like a train in the distance or the eidetic image of a just prior percept, takes on the attributes of a memory or a perception that is uncertain. It is not a question of a past or present object. *The degree of revival establishes the pastness or presentness of the perceptual object.* When perception is replaced by hallucination as in dream or the hallucinosis of waking psychotics, the hallucinatory world becomes a world of present objects. One needs an actual world for an image to be past in relation to. The world of perception establishes a *now* in relation to which an incomplete perception takes on a quality of pastness. Without a veridical world, as in dream, the image world becomes the present of the observer, in relation to which only the most rudimentary past can develop.[12]

Clinical studies are filled with examples of transitional cases. For example, there is a clear transition from a percept to an eidetic image to a memory image (Klüver, 1933) in which the subjectivity, the veridicality and the pastness of the image change as a whole in a withdrawal back through the structure of the original percept. Such examples confirm that the past of memory and the now of perception – persistence and actuality – are momentary aspects of percept-development or, what amounts to the same thing, stages in the realization of a memory into the present.

Self and introspection

The *Project* is notable for the attempt to describe stages in cognition in relation to *Cs*, though the nature of *Cs* itself and the sources of *Cs* activity are scarcely examined. The remainder of this chapter explores the implication of microgenesis for a theory of *Cs*.

The emergence of a self is linked to an enhancement at penultimate phases in the microgenetic traversal. The self is intermediate between the potentiality of memory and the now of perceptual experience.[13] The mnemic source of the self contributes a feeling of identity and persistence 'over time'; the perceptual surface is a 'screen' for the self to scrutinize. The intermediate position gives a *Cs* of the evanescent images, and still more fleeting (in perception, if not the material world) external objects, into which it – the self – distributes. The result is a kind of theater in which the self gazes on the images, the objects and the space, of its own creation. The model requires a subjectivist perspective in which objects are understood as exteriorized mental images.

A core self generates a *Cs* self,[14] then the space of the imagination, which in turn leads to external objects. An object world is a necessary ground or contrast for introspection, which is the state of a self conscious of a proposition, a concept or an image. *Cs* is the relation of self to image or object. The relation of self to image (introspection) is nested in the bridge from the *Ucs* to the external world (Figure 4.3). Introspection obtains when a self is aware of a state of *Cs* even if the contents of the state are external objects. The awareness of the state involves an awareness of concepts in addition to objects. Since a concept is a precursor of an object, a heightened state of self is a prominence of pre-processing phases in object formation, thus the appearance of conceptual or imaginal content.[15]

The formative direction is from self to world, with *Cs* anchored in the self. *Cs* is the proximate source of the self, the world its outer limit. If an object world is not achieved, objects no longer exist and *Cs* is that of dream. The *Ucs* at one end, perceptual objects at the other, are the

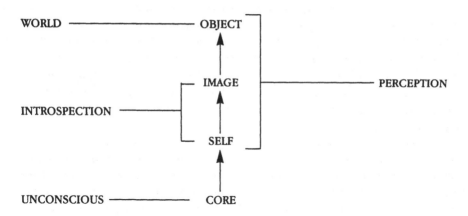

Figure 4.3. In the evolution of the human mind/brain there is an expansion of a segment in the process that lays down the external world. This gives rise to the self and the private space of images. The appearance of this phase is the basis for introspection which is the relation of self to image. This relation is embedded in the relation of a subject to an object, that is perception. A self-Cs perception, a perception that is more than animal awareness, requires an implicit engagement of introspective content.[16]

'physical' boundaries of *Cs*. The transition from the *Ucs* (core) self to the *Cs* self is the internal boundary of *Cs*. The external boundary is the transition from an image, including the body image, to an 'independent' object.

Consciousness

In the *Project*, the contents of *Cs* are sensory qualities that are represented by different periods of neuronal motion. *Cs* looks outward at peripheral sensations and inward at images and ideas. Freud wavered between the view that *Cs* was a mere accessory or 'inseparable from the physiological mental process', concluding vaguely that 'consciousness is the subjective side of one part of the physical processes in the nervous system . . . (but its omission) does not leave psychical processes unaltered' (SE 1, p. 311). I would say that *Cs* inheres in the nature of its content, that *Cs* is not a function but a momentary achievement, one that with its content must be continuously renewed.

We speak of a conscious self, but does *Cs* (*Cs*-of) not require a self, and is the self not the agent or the recipient of every conscious state? If the self were not conscious, the state would not be a *Cs* state minus the property of *Cs*, but would express a different state and a different (*Ucs*, core) self. The self and state of *Cs* do not fluctuate independently. An alteration in *Cs* entails an alteration of self. *Cs* does not change and leave the self unaffected. The self cannot be lost and the object world cannot disappear without a profound effect on *Cs*. The interdependence of self, *Cs*, and objects is clear in pathology.[17]

If this is correct, *Cs* could not be an appendage or property of a particular type of mental state, for one could have the mental state unchanged
with the property extracted. A bird shaved of its feathers is still a bird. If *Cs*
is a property, it is an essential property such as 'living thing' is to bird,
without which the entity ceases to be that which it is. *Ucs* mentation is not
conscious mentation with *Cs* removed. If *Cs* were a property, the taking of
Cs for a mental state would substitute an attribute for the entity it
modifies.[18]

The opposite approach is to conceive of the mind as the equivalent of
consciousness, in which case the concept of mind hinges on the quality of
conscious phenomena and their limits: quality with respect to the nature
and scope of subjectivity, i.e. what species of thought or behavior merits
inclusion in mentality; and limits, with respect to whether the mind incorporates *interior* non-conscious states including *Ucs* ideas or *exterior*
contents in perception. The first question is, what segments of cognition
and the world are conceptual.

The conceptual approaches the physical at both ends in a continuum:
at one end a core that elaborates *Cs*; at the other its surface in the external
world. Some writers include the *Ucs* as part of the 'furniture of the mind'
and exclude external objects. Others draw the line at objects, i.e. the mind
is perspectival but objects are not perspectives. Still others include
external objects as contents in *Cs* but exclude *Ucs* cognition. If *Cs* is
identical with mind, *Ucs* mental states do not exist. The terms *Ucs* and
mental would be incompatible, though *Ucs* states could in principle
become mental – or retrospectively mental – by virtue of becoming
conscious.

While *Ucs* mentation and perceptual objects may or may not be
included in the mental, many assume that *Cs* cognition and perceptual
objects reduce to material brain and world process. An *Ucs* urge corresponds with the discharge of 'drive' centers in the brain, a perceptual
object corresponds with a physical entity in the world or the material
substrate of that entity in the brain. The question is whether the substrate
of *Cs* can be understood if mind does not extend to the world 'beneath'
awareness or the world 'around' it.

The world does not pass from the physical to the conceptual, or the
reverse, but is either physical and/or conceptual all the way through. If one
object in the world is conceptual, all objects are: tables, unicorns, quarks,
all existing or conceivable objects. A table is a conceptual object at the
mind's outer limits, a distal segment of the conceptual realized into the
world. Or, it is a realization by nature of 'facts' in the world through the
agency of a given observer. The world represents itself in innumerable
versions of what the physical might be like. The potential world of mind,
and perhaps of nature, is infinitely richer in possibility than is the actual
world in fact. That is because only a portion of its objects actualize.

Unconscious mind?

Phenomena postulated by psychoanalytic theory such as repressed contents, *Pcs* thoughts or *Ucs* conflicts are inferences about experiences at a depth when conscious behavior can not otherwise be explained. The fact that such experiences appear on a continuum with conscious mentation argues for their inclusion in a wider account of the mental. *Cs* waxes and wanes over the course of the day. Some events are in focal awareness, others are in the background. A variety of intermediate phenomena, dream, reverie, drowsiness, distraction, confusion and pure object awareness, as well as borderline states, subliminal, ambient and other *Ucs* perceptions, and a variety of behaviors with all the trimmings of mentality except the ability to describe them, ranging from experimental studies of semantic priming to the popular literature on hypnotic states, the responses of children, animals and the 'isolated' right hemisphere, are evidence that the mind does not cleave into *Cs* and *Ucs* but is a continuum from inaccessible states of primordial cognition through dream and the barely conscious, to object awareness, states of meditation, and intense concentration.[19]

In addition to the gradations from *Cs* to *Ucs*, there are a variety of states of *Ucs* cognition, from 'deep' sleep to 'transitional' phenomena, including the different forms of dream, nightmare, hypnagogic and hypnopompic states, sleep terror, somnambulism, sleep-talking (in non-REM sleep), and so on.[20]

The argument from the standpoint of *Cs* (i.e., the equating of mind with *Cs*) obviates the difficulty in describing a non-conscious state of mind (there are none) and eliminates the problem of how the non-conscious can have mental properties (it can't). Searle (1992) has argued that an *Ucs* mental state implies an accessibility to *Cs*. Without access to *Cs*, *Ucs* experiences or ideas do not exist as *mental* phenomena. He writes, 'we have no notion of the unconscious except as that which is potentially accessible.'[21]

How would one determine potential for access other than by an access that is successful, or a description modeled after conscious events? Aside from the evidence (above) for *Ucs* cognition, this leaves out those transitional states in which some form of *Cs* awareness is present. Searle recognizes this difficulty for elsewhere he writes that 'a system is either conscious or it isn't, but within the field of consciousness there is a range of states of intensity ranging from drowsiness to full awareness' (Searle, 1995). This seems to be a theory of *Cs* as a searchlight with an adjustable dimmer. When the intensities recede beneath a margin of accessibility the state becomes *Ucs*. While the idea of gradations allows the incorporation of states of incipient *Cs* as well as stages of progressive *mental* growth in evolution, the concept of intensity turns the state into a quantity when, in fact, every gradation is a qualitative shift. The self of the dream is a passive

onlooker, even a victim, to dream events, not a lower intensity of the
agentive self of wakefulness. The fluid space and melting images of dream
hallucination are features of another world, not a lesser degree of the
empty space and stable objects of waking percepts.

Intentionality and *Cs*

Many philosophers, especially Ryle and the behaviorists, have attempted
to divest cognition of *Cs* entirely so that *Cs* and *Ucs* states, or behaviors of a
'*Cs* and *Ucs* type', are judged to be comparable regardless of the nature of
the accompanying inner experience. The argument that human intelli-
gence can be reproduced in a machine depends on this divestment. It is
easier to eliminate *Cs* from the human mind than to assert it in the
workings of a machine. In recent years, intentionality has been the battle-
ground on which this move is fought.[22]

From an external perspective, that is, observing behavior from outside,
the conflation of the purposeful with the volitional makes goal-direction
the criterion for an attribution of intentionality. If the organism displays
goal-directed and purposeful behavior, such behavior is sufficient to infer
an intentional state. Those who use this strategy may give lip-service to the
elusive quality of private experience, even as their claims for sufficiency in
the judgment of what is intentional by an observer trivialize the import of
Cs and its internal architecture. The mental state is much richer than the
behavior it deposits, yet that richness is collapsed, along with the behavior,
onto the hollow promise of an explanation by a neuroscience of the
future, when neither mind, behavior, nor brain state have been examined
with the care they deserve.

The purposeful is not identical to the volitional, as sleepwalking is not
the same as walking when awake. The somnambulist has neither the option
of not walking nor a menu of alternative acts. Purposefulness can follow on
the heels of a decision but it can also usurp an act of choice. An action
develops from a potential that is equivalent to a state of implicit indecision.
However, the potential choices in the background of a decision are not a
multiplicity of options waiting to be selected but share features with a
configuration that has the potential to develop along several paths out of
which one option is realized with clarity. Choice requires minimally a state
of indecision, uncertainty or conflict. If choice is the prominence of an
embedded phase in purposeful action, the shift in relations through which
this phase becomes prominent coincides with the emergence of an introspective awareness as an enhancement of earlier stages in microgenesis.
Indeed, the continuum from dream awareness (*Ucs*) to object or activity
awareness[23] to self-awareness (introspection) corresponds with the
sequence from the automatic to the purposeful to the volitional.

The purposeful and the volitional, therefore, refer to different sets of
relations in the derivation of action and choice. A dog that digs up a bone

exhibits purposeful action but no one would contend that the dog contemplates an absent bone, or entertains a choice as to whether or not to dig it up. The choice that is implicit in purposefulness becomes explicit in volition. This 'becoming explicit' is the essence of *Cs*, and accompanies other phenomena related to an accentuation at preliminary phases. In volition, choice is in the foreground of awareness, that is, choice in the mind, not in the world.

The coming-to-the-fore of concepts and images as intentional objects within an introspective state of *Cs* depends on the same set of relations that determine the prominence of other mental contents. For example, in a manner similar to the microgeny of choice, there is a parallel development from instinct to drive to desire to discriminated states of feeling in the individuation of affect (Brown, 1996, pp. 139–154). The intentional feelings of *Cs* desire and preference emerge out of *Ucs* drives through a process of fractionation. This process is guided by intrinsic and extrinsic constraints that determine what configurations individuate to awareness. The hypothesis of repression or censorship is the turning into a mechanism of the exercise of these constraints on the microgeny of images and affects. Those configurations that survive are shaped to fit the environment, as contents develop from the drive-based, thus 'selfish', core of the mind to the reality-based, thus adaptive, surface of the world.

From *Ucs* to *Cs*

In process theory, contents in *Cs* are not the result of a bringing to the surface of what is submerged, but the effects of level-specific constraints on what eventually actualizes as *Cs*. There are several paths available to *Ucs* cognition: (1) it can actualize directly, as in dream, when the microgeny of the mental state is attenuated, or develop along endogenous lines without the constraints of sensation. The knowledge of such states, because they are *Ucs*, depends on those elements that are derived to *Cs*; (2) it can discharge into *Cs* cognition, as in hallucination or psychotic ideation, in which case the awareness is occasioned by an occurrent *Cs* state into which the *Ucs* material penetrates. The intrusion of *Ucs* material into *Cs*, for example as in the case of a persistence of dream elements on waking, permits a glimpse of the extent to which *Ucs* cognition deviates from that of *Cs*; (3) it can undergo qualitative transformation by configural effects on the developing configuration as it passes into and generates *Cs*. This effect is the primary manner in which *Ucs* material becomes *Cs*.

This implies that the *Ucs* is not known through the *transfer* of content to *Cs* or by a *Cs* attachment to its content, but through configural effects on the becoming of states of the incipient present. Every content has a momentary history in which *Ucs* and *Pcs* phases are embedded in – actually transformed to – an ensuing segment of process (becoming). *Cs* is not a copy of the *Ucs* or a derivation of replicated elements. but an outcome of *Ucs* effects on the to-be-*Cs* content. This is consistent with Freud's depiction of *Ucs* mechanisms as differing from those of *Cs* but it is inconsistent with a single trace or re-representation model (see above).

If *Cs* is caused by or rests on non-conscious activity, the better part of the state of *Cs* is ignored by excluding the non-conscious portion. If constraints are construed as causes, an *Ucs* state that transforms to a *Cs* experience might be said to have a causal role in its induction. If *Ucs* content (or its brain state correlate) is the cause of *Cs* content, the *Ucs* content would be manifest as a prior cause. If the *Ucs* is interpreted as nothing more than brain activity, and causal with respect to *Cs*, whether *Cs* is also just brain activity or entails something more, a series of states between *Ucs* and *Cs* cognition, or from an *Ucs* brain event to its ensuing *Cs* (mental) correlate, is required in the 'causal chain' from *Ucs* to *Cs* unless the penultimate *Ucs* step is the ground of an emergence to the mental.

If the *Ucs* is causative, or serves as the ground of an emergent step, i.e., if *Cs* develops out of a preceding *Ucs* state through a causal or emergent process, the identification of mind with *Cs* focuses on the endpoint of the series and severs the immediate causes or anticipatory stages of *Cs* from their effects. *Cs* is then uncoupled from its antecedent states and set adrift like a raft cut loose from its moorings. The disjunction of *Cs* and *Ucs* cognition mutilates the temporal relations that underlie the appearance of *Cs* and shatters the *process* through which such states actualize, i.e., the becoming that binds the formative moments of the mental state together.

On this view, the incorporation of an *Ucs* or *Pcs* core and a *Cs* self into a description of mind expands the internal boundaries of the mental just as the incorporation of acts and objects expands its external boundaries. The expansion is necessary. *Cs* depends on intrinsic relations between segments in the mental state so the boundaries of the mental state are critical to how these relations are conceived. In microgenetic theory, the mental state does not stop at the eyes or ears but includes the worlds perceived by those organs. Core and world are its limits.

Bradley wrote, 'every relation essentially penetrates the being of its terms'. This is definitive with respect to *Cs*, which does not exist as an activity or mechanism apart from the relations that underlie its appearance. These relations arise in the temporal flow of mental process. A fuller understanding of *Cs* requires a sensitivity to these relations, to the precedence of moments in becoming and, especially, the contrast between the duration of the present and the immediacy of physical passage. Mind is not a succession of instantaneous states that seems to trail off in memory like the still-present notes of a melody, but obtains in the layers of felt transition through which each note, each *Cs* present, actualizes.

Chapter 5
The Unconscious (Freud) and Process Theory

The foundational concepts of psychoanalytic theory are elusive. One reason for this is that there is more than one theory. Another is that the 'standard' theory is not just the set of Freud's writings, or the consensus interpretation, or the most recent opinions on their meaning, but a mix of Freud's changing views with those of his interpreters, along with the practical concerns of the clinicians. This would not be problematic were the implications of antecedent versions annulled by the ensuing ones, or if some 'bottom-line' metapsychology could be extracted on which there was wide agreement. Instead, the permutations of the theory were asserted, abandoned, then reasserted, while the collective edifice, remaining more or less intact, underwent a gradual accretion, like the stacking of unconscious memories, regardless of the fate of the under-lying assumptions. It is surprising to see incompatible statements coexist as allowable perspectives, or harmless quibbles over the exegesis of unsupportable assumptions take the appearance of serious disputation, while the metapsychology on which the claims of clinical theory rest seem strangely irrelevant to the entire enterprise. Clearly, psychoanalysis is more protean and resilient than any of its foundational precepts.

Still, in spite of many reservations, I believe there are postulates common to the successive models that can provide a framework for a subjectivist theory of the mind. These include certain of its microgenetic implications, the topographic and genetic concepts (Schilder, 1953; Rapaport, 1967; hereafter cited as R[1]), dreamwork and paralogic as mechanisms of primitive ideation, the parcellation of instinct to feeling, the relation of normal cognition to the process to symptom formation, the ethological principles, e.g. the instinctual core and innate origin of psychological phenomena, and their epigenetic unfolding. In this critique, it is my intention to compare some of these basic concepts to those of process metapsychology and the microgenetic account of the mind/brain state.

Unconscious cognition

Unconscious processes are, presumably, the causes of conscious experience, but the evidence for this – from everyday experience to hypnosis, dreams and the neuroses – consists of the products of unconscious activity, not the activity itself, though Freud claimed a more limited and direct connection, i.e. 'unconscious processes only become cognizable by us under the condition of dreaming and of neuroses' (Freud, 1953 (hereafter cited as SE); 14: 187). These effects cannot be self-caused, i.e. a dream cannot be both the cause and effect of its content. Since we can only observe effects, they must, it would seem, be the outcomes of antecedent processes, possibly of a different type. Moreover, the expressions of unconscious activity are filtered through conscious verbal reports – language, for many, is essential for consciousness; for Freud, it is a means of access – so that the causation of a conscious behavior, or even dream consciousness, is, as analysts would surely agree, an inference on the nature of upstream process.

In psychoanalysis, dream and perception are phenomena with very different explanations, while in microgenetic theory they are endpoints at different phases in a single process. Suppose we construe a dream or an hallucination as a perceptual equivalent that actualizes prematurely. The dream represents a subjective phase, the object, an objective phase. On this view, a waking hallucination is a submerged fragment that intrudes in a perception. In dream, the phase to which the fragment refers generalizes across the modalities to become another reality. From the standpoint of either theory, the objects of dream and hallucination are comparable to those of a perception but antecedent to them. We would agree that the perception of an object world is the outcome of a process that does not give the process that perceives it. We do not assume that the objects and relations in a normal perception are ingredient in the process that generates them. If we cannot adduce the *process* of percept formation from the content of the perception, why should we expect to derive the unconscious process of the dream from its content?[2]

The symptoms of unconscious activity are the transformed products of an activity that is unobservable. Can these symptoms then be a guide to the infrastructure of the causal process or are they better conceived as justifications or reasons for the existence of the unconscious? Among others, Wittgenstein noted the confusion in psychoanalytic writings of reasons and causes, the former requiring assent, the latter scientific documentation.[3] If the metapsychology cannot withstand close scrutiny, this is not to say that Freud's account of the unconscious is wrong, only that its evidentiary basis is inconclusive. The point is that an account of the unconscious should not rest on the material of psychoanalysis alone but should seek additional sources of support.

The mechanisms or properties of the unconscious include the primary process, dreamwork, drive-representations, emotion, infantile reactions, timelessness and an absence of contradiction.[4] Besides the idea and the memory trace, which are the 'bricks' of the metapsychology, and the drives and connections that are its 'mortar', what else is in the unconscious? In one sense, everything concealed to the subject of his own brain activity is by definition unconscious. Yet we do not consider brainstem reflexes unconscious, nor cerebral mechanisms such as those for eye movements, color coding, feature detection, and the like, or endocrine systems. Nor the part-acts that go into a complex behavior, like walking, nor some implicit processes such as after-images and constancies. The conceptual architecture of the mind, some habits of thought, lexical-semantic fields, possibly some pragmatic aspects of speech, might be included, but not, presumably, the prosodic, phonetic and articulatory rules or regularities that govern speech production. Perhaps it is a matter of how 'hard-wired' the memory is, or the 'degrees of freedom' in its operation. Memories that are latently psychical are part of the unconscious. According to Gill, 'Freud regarded all that is dynamically unconscious as part of the system *Ucs*', though in later writings, the *Ucs* became restricted to the repressed (Gill, 1963 (hereafter cited as G), p. 23). Schilder wrote of the 'disappeared parts' of learned skills. He distinguished organic or structural formations from the unconscious proper.

At times, Freud treats contents or mechanisms in the unconscious as qualitatively distinct from those in consciousness, for example, the claim that instinctual drive acts only in the unconscious, or that unconscious cognition is primitive, or infantile, or what is normal for a child but no longer dominant in the adult, e.g. unconscious contents are like 'an aboriginal population in the mind . . . (plus) what is discarded during childhood development as unserviceable' (SE 14: 195). The infantile character of unconscious ideas seems to work its effect largely in the unconscious, inciting conflict and exclusion of unacceptible ideas from consciousness. Presumably, the ideas become less infantile as they surface, but how this occurs is a mystery.

More often, Freud emphasizes that an idea in the unconscious is the same as when it becomes conscious, or that the same idea is re-registered in consciousness, or that the idea, if not re-registered, can take on a conscious quality. In other words, when an idea in the unconscious becomes conscious, the censorship permitting, there is either a fresh record, a second registration in consciousness, or a change in the functional state of the same material in the same locality. In all cases, what has changed is not necessarily the content, but its level of activation or availability to consciousness. Freud wrote, 'the only respect in which (some latent states) differ from conscious ones is precisely in the absence of consciousness'.

Freud described the unconscious and the process through which unconscious ideas become conscious without addressing the distinction of the cognitively inert from the cognitively active, nor the transition from a primarily neural event to one that is primarily psychic. The unconscious consists of those contents which have the potential to become conscious, give rise to conscious symptoms or influence conscious outcomes. The assumption of a transfer from an unconscious to a conscious compartment recalls the re-representation of Hughlings Jackson. This concept was replaced by an activation model that did not necessitate a relocation of the trace, though it did entail a searchlight mechanism for consciousness. The concept of a qualitative transformation was an emendation by later writers.

Unconscious content

In psychoanalysis, the idea is an atomic element with causal properties that is besieged by positive and negative forces.[5] Gill (1963) wrote, 'When the *Ucs* process has been inhibited, it has become a *Pcs* process. It has not undergone a distortion of *content*, as has a process which has been subjected to the counterforce of defense or censorship.' Since the censorship is a counteractive energy, this implies that inhibitory but not excitatory energy alters content, and that an alteration of content is a 'distortion'. Except for the distortion induced by the countercathectic charge, an idea is not obligated to change in the transit to consciousness. Repressed ideas drop out, but those that survive are not necessarily altered on becoming conscious. What changes is the *presence* of consciousness. There may also be a recombination of ideational primitives to more complex entities. The aggregation of simple ideas to complexes of ideas, i.e. the expansion to wider fields of traces, though not explicitly argued by Freud, could serve to

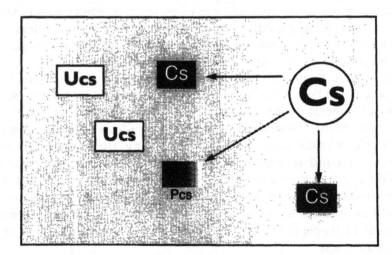

Figure 5.1. On the functional or dynamic theory, consciousness arouses traces directly to *Pcs* or *Cs* statues.

heighten the cathectic charge of the complex. Those portions of a complex idea that are not acceptable to consciousness may be repressed, while other portions become conscious directly.

Whether an idea undergoes a change in registration or functional state, it remains, for Freud, the same idea, either relocated or empowered. We find this interpretation in neuropsychological writings. Thus, Kurt Goldstein wrote that 'the same phenomenon, such as an idea, an act of will, a feeling, a habit, etc., can be conscious at one time and unconscious at another'. Of course, when the idea is unconscious, we have no way of knowing what the idea is. That is why organic symptoms are valuable, because they disclose subsurface process. In this respect, the copy or translation theory of psychoanalysis differs from the microgenetic concept of a conceptual and semantic transformation of primitive constructs.

In contrast, in microgenesis the configuration develops out of a dream-like cognition and personal memory into perceptual reality. There is no actual content that is vulnerable to distortion, i.e. the distortion is not secondary but typifies the phase. In fact, there is no content that is acted upon; rather, a configural pattern becomes a content when it discharges (actualizes). Otherwise, it serves as a contentless potential for an ensuing transformation. Specifically, a content is whatever actualizes when the process completes itself. Moreover, preliminary phases share in the final content even as they are transformed to it. The entire configural wave, from the unconscious to consciousness, is ingredient in the final actuality.

In this respect, there are similarities to the topographic concept that when an idea becomes conscious, its position in the unconscious system is probably retained, since the idea still exerts an effect in the unconscious after it has become conscious. However, the continuum from unconscious to consciousness in microgenesis is a qualitative shift, whereas in psycho-analysis, at least in the early model, the *same* content can have a separate location in conscious and unconscious systems. This implies a multiple representation of the same idea. Freud abandoned this concept, but it resonates with the modern, in my view, incorrect, interpretation of 'retrieval' as a 'looking up' of contents in a memory 'file' and bringing them to consciousness. The concept of multiple representations is also consistent with connectionist and holographic principles.[6]

Among analysts, the more prevalent view is that an unconscious or preconscious idea can undergo a change in functional state and become conscious without a shift in its location. On this account, consciousness is like a scanner with the idea its target, or the idea may act to draw cathexis from consciousness. The system *Cs*, like that of *Pcpt.* (perception) with which it was closely identified, does not contain content, but was conceived as a 'sense organ' that must be stimulated for a mental content to become conscious (G 62). The sense organ can incite a content or be attracted to it. Freud wrote that 'a train of thought cathected by some aim becomes able under certain conditions to *attract* [my italics] the attention of *Cs.*' This is how hallucinations are said to become conscious. Dream consciousness is the dream plus the consciousness it attracts. The notion that *Cs* is a sense-organ that attaches to ideas preserves the energic theory, and all its difficulties, in the view that conscious attention is a kind of higher-order purposive cathexis. In contrast, the possibility that hallucination *generates* a fragment of a dreamlike consciousness which, in dream, expands to all modalities, does not appear to have been given serious attention.[7]

In any event, whether propelled into consciousness, attracted by consciousness, or accompanied by a consciousness specific to the degree of perceptual realization, the idea is conceived as an isolated element which can be selectively tuned, with distortion arising from traces associated to it, not through internal change, i.e. the distortion arises in the substitution of one idea for another, not as an intrinsic property of the unconscious trace. The rich narrative of clinical analysis that constitutes a symptom-complex, for example an oedipal saga or dream interpretation, is far removed from its explanation in traces that are energized by electric currents. In my view, the dreamwork is an account of the process of normal thinking, a preliminary and obligatory phase for all conscious content. The spatial and temporal dissociations, or reconnections, of the trace or the idea, suggest but do not support a theory of contextual relatedness and the nesting or embedding of categories in terms of affective and experiential relations.

In dynamic theory, conscious, preconscious and unconscious are properties of ideas, not the compartments in which they are located. Ideas activated into consciousness are entrained in a momentary assembly regardless of their 'location'. In this model, Freud relinquished the Jacksonian framework of levels in favor of the modern notion of a kind of network in which psychoneural elements (traces or ideas) participate in conscious or unconscious activity irrespective of their location. The elements are activated into a widely distributed circuit depending on their relatedness to drive and their degree of cathexis. It is an odd conclusion of this theory, and one reason for its abandonment, that the topographic level of an idea (conscious, preconscious, unconscious) does not determine whether the idea is conscious without relocating to another component.

The shift to a functional theory was occasioned by the discovery that the unconscious is not uniformly repressed, while the ego is not wholly conscious, so that conscious, preconscious and unconscious referred to the level of activation of a sessile trace, e.g. 'what we regard as mobile is not the physical structure itself but its innervation' (SE 5: 610–611). The claim that ideas were static and that the cathexes were the mobile elements – almost counter-intuitive, given the plasticity of ideas in relation to the fixed repertoire of the drives – was the basis for the dynamic theory of id, ego and superego. When Freud discarded the topographic model of unconscious and conscious systems for a dynamic account of states of ideas, the bond with neuroanatomy, both real and speculative, was shattered. The ego or superego are mental agencies which do not have known or surmised correlates to anatomy. An explanation of behavior by reference to these agencies distanced the model from neural concepts and paved the way for the free-flight of psychoanalytic theory.

What is an unconscious idea?

As is well known, the concept of an unconscious, and of instincts and ideas in the unconscious, preceded Freud's description of the fine structure of the *Ucs* and provided the climate in which his theory could take root. The notion of an unconscious ideation driven by the power of instinct was strongly imprinted in the philosophy of the day. The spatial metaphor of unconscious ideas as entities in a separate realm of the mental life, as in the depiction by Fechner of a sphere of 'ideas below the threshold of consciousness', was probably derived from Herbart through Freud's teacher in philosophy, Brentano (MacIntyre, 1958). Leibniz suggested that unconscious ideas were too weak to enter consciousness, while Hartmann (1868; hereafter cited as PU) observed that this explanation was inconsistent with the power of unconscious ideation. It remained for Freud to resolve the matter by proposing that the ideas drew their power and persistence from unconscious drive energy, yet were excluded from consciousness by the countercathexes and the layers of defense.

What does Freud mean by an idea, conscious or unconscious? An unconscious idea is both prelinguistic and preconceptual. Since the idea derives from the trace, which in turn derives from the perception, one can ask, how does the idea take on propositional content other than by association to verbal memories? With regard to the trace, it first must undergo a cathexis, a libidinal charge, or more simply, an activation. Once a trace is activated, it is no longer a memory, it becomes an idea. A memory trace is an inactive idea, an idea is an activated memory. Freud wrote that the only difference between a trace and an idea is the cathectic charge. The trace is a dormant idea; indeed, it does not really exist prior to its activation – it is the *potential* to be activated. Nor does the trace exist even when it is activated, for then it becomes an idea. A trace that is inactive, as in the description of competence in linguistics, is not something that actually exists. An inactive trace corresponds to the relative strength of connectivity in a virtual pattern of cells, a pattern that exists only as the potential to become an idea. A store of 'virtual knowledge' in the unconscious was first proposed by Leibniz, and restated by Hartmann, who emphasized the timeless quality of unconscious ideas.[8]

The drive activation organizes the ideas according to its needs. An idea, then, is an activated memory or a combination of a memory with drive energy, a 'drive-representation'. At this point, the idea is still unconscious. Freud had little more to say about ideas and, because they were assumed to be basic entities, thus unanalysable, nothing to say of their internal structure. And not just ideas. Many authors have noted the poverty of Freud's discussion, as Strachey put it, of 'some of the *most* fundamental of the concepts of which he makes use: such concepts, for instance, as 'mental energy', 'sums of excitation', 'cathexis', 'quantity', 'quality', 'intensity', and so on' (SE 4: *xvi*).

In Freud's theory, a trace is comparable to a note on a sheet of music, a code that is a prompt to the production of a tone. Once a trace is activated by the energy of the drives, or a note, say, by the vibration of the strings of an instrument, it passes to actuality from virtual non-existence. At the point of its arising, the trace and the idea already differ in temporal orientation. The trace looks backward, as far into the past as the perception from which it issued. The idea looks forward to an action or an object. A trace that is active is no longer a memory but a direction to the future. Is this the effect of libidinal charge? Drive is directed to the future, or to an object in the future of the drive. The sense of the future is the surge to the actual in each cycle of the mental state.

Affect and activation

Rapaport writes that the idea becomes conscious when a surplus of cathexis, a hypercathexis or an attention-cathexis of secondary process, becomes attached to an idea or memory trace. The libidinal nature of the activation, the motivated or wishful quality that determines which traces will be activated, is central to psychoanalytic formulations. However, actions are generally motivated by self-interest. The self is primary because the genes and their expressions in the drives are selfish. But behavior is also self-interested because the intensity of the drives is maximal intrapsychically, and mitigates, microgenetically, as the affect distributes into objects. The microgeny of affect is from drive through feeling to value as concepts objectify. The trajectory of the object in its development from mind to world is accompanied by a specification of drive to the valuation that animates the least perceptual entity. In this development, the self is the center of its own affective field. An object, a rock or another person, receives a share of the conceptual feeling permeating the subjective phase of the microgeny, which weakens from drive to value as it is apportioned to the perceptual field of the observer.

Put differently, the affective tonality of an external object is a product of the observer, who transfers feeling in the form of value into the object as it individuates. According to microgenetic theory, the possibility of human altruism, or of actions not motivated by self-interest, depends on the retained valuation in the self of the feeling it generates for other objects. There is a gradation of intimacy from the inorganic and the indifferent to the familiar and the beloved that reflects the intensity of the subjectivity that is invested in an object development. This can reach the point – say, a morbid despondency over the loss of a loved object – where the value of the other is, to the observer, greater then that of the self, or the self-concept may be defined by the valuation assigned to the other, presumably because the valuation of an object (or idea) arises in the valuation of constructs composing the self-concept.[9]

Selectivity

How are we to understand the neuronal basis of drive energy and the cathexis of the trace? How does a drive, prior to concepts and memories, conceived as having its onset in instinctual energy, or even restlessness, anterior to the appearance of form, signal the object it is seeking? Does drive energy circulate among neurons, or populations of neurons, that correspond to the trace? Should the energy be conceived as a state of tonus, a current, a lowered synaptic threshold, or a patterned configuration that incorporates a set of traces? Freud, writing prior to the definite refutation of the syncytial or reticular model of nerve conduction, but clearly, as early as the *Project*, on the side of the neuronal theory,[10] said that drive-cathexis 'is spread over the memory traces of ideas somewhat as an electric charge is spread over the surface of a body' (SE 3: 60).

The drive calls up those traces that prefigure the object toward which it is directed. A generalized change in threshold, or the spread of an electric discharge, or the differing degree of facilitation of greater and lesser quantities of energy, does not give the specificity of trace-selection to the need-satisfying object, yet no rationale is given for a local change in threshold that could give selective activation.

The selection of the trace is no less problematic than the attachment of drive energy, which is central to the theory. Does the trace have an affective component, is it stripped of feeling in order to receive the libidinal charge, are events 'stored' and evoked independent of the feelings that accompanied them, is a cathexis merged with a feeling tone in the trace? Once the trace becomes an idea, does it *contain* feeling in the form of drive energy, i.e. are ideas penetrated by feeling, or are they neutral entities to which energy is applied as an impulse to arousal? Conversely, do unconscious feelings take on ideational content when they attach to memories? Freud proposed that instinct sorted through the traces with its object as a guide. Trial and error, and a facilitation of synapses, are a species of explanation but, as Hartmann noted much earlier, they do not account for the *specificity* of the unconscious influence on consciousness, e.g. in wit, in directed thought, etc. (PU 1: 303), nor is it clear how learning alters a drive object that is unconscious. Moreover, ethological studies demonstrate that learning, at least at the level of instinct, does not so much provide as *fine tune* the efficacy of a figural gestalt.

Rapaport says the instinctual wish strives to attach itself to memory traces suitable to its expression. What could this mean? In psychoanalytic theory, drive cathexis is a form of attention with a limited amount of energy. The ideas attached to the cathexis consume this energy, which is then unavailable for other ideas. Before Freud, the question was, how much energy is necessary to bring an idea to consciousness. After Freud, the question became, or should have become, how does the energy find, attach to, and organize traces appropriate to an instinctual goal? How does a *mobile* energy that is comparable to an electric charge locate its targets, or the targets lure their cathexes? An expansion of synaptic excitation or inhibition to cell populations and psychic entities still does not explain their specificity and goal direction. Freud was aware of the disjunction of a quantitative theory of drive with the quality of mentation, especially in consciousness (SE 1: 309), and offered an explanation on a neuronal basis relating to permeability (and periodicity). If the pure, unattached drive energies differ at the start, is the drive conditioned for a particular trace, does it modify the trace, or undergo further specification on attachment? According to Rapaport, Freud did not clarify these concepts.

Freud distinguished two aspects to a drive-representation, one qualitative (ideational), the other quantitative (energic). Instinct is a quantity. So is restlessness. Freud wrote, 'an instinct is without quality'. Does quantity map to quality? The problem of selectivity in activation, as in decathexis (see below), is a fatal weakness in a theory that begins with cognitive singletons, intrinsically opaque, that act by way of external relations. For Freud, an unconscious idea is a copy of the idea in consciousness, less the distortion of contents that are ordinarily shunted, by repression, into other paths. But selectivity is the *goal* of an act of thought, it is achieved by the *individuation* of mnestic, imaginal-perceptual, affective and linguistic content. The multiplicity of the mind or the world is not obtained by a correspondence of unconscious to conscious ideas, or of conscious ideas to world objects. It is achieved in a process of specification through which an inner and outer world of adaptation individuates out of unconscious unity.

Clearly, these objections stem from the solidification of abstracts and their artificial bifurcations. Gudmund Smith has remarked on the persistent associative logic in Freud's work, for example, the reference to 'mechanisms' (defense), and the lack of interest in gestalt psychology. We see this logic at work in the opposition of drive and idea, memory and perception, cathexis and anti-cathexis. Entities are reunited through external bonds that should never have been separated in the first place. In this respect, Freud took a conservative position. Reason, ideation, logic, thought, etc. have been opposed to emotion, appetite, instinct and will since Plato. On the other hand, Hartmann was closer to the truth when he wrote that idea and will, i.e. trace and drive, were an 'inseperable unity', and that consciousness and intellect depend on 'the emancipation of the idea from the will' (PU 2: 83). The liberation of thought from emotion is the function of consciousness. Indeed, while consciousness is essential for the individuation of drive and idea, their attachment, in psychoanalytic theory, is essential for consciousness. Why would the marriage of drive and idea that is necessary for consciousness lead to a sudden divorce once consciousness is achieved?

The microgenetic interpretation is that the conceptual and affective are combined *at the start*, with fractionation of *conceptual feeling* into partial affects and object concepts that distribute, finally, into the value-laden images of the objective world. A phase of primitive constructs invested with drive energy, tendencies or potentials for a goal rather than activities containing one, transition to affectively-charged acts and objects through a graded parcellation of conceptual feeling. Schilder described a progression to partial affects and increasing definiteness from an early phase of affect-laden imagery. The sequence is similar to Rapaport's account of a diffuse global image of a need-satisfying object that differentiates into discrete objects connected by threads of feeling to the drives. On this interpretation, memory and drive, idea and feeling, value and object, are the changing manifestations of successive phases as the mental state undergoes qualitative transformation *en route* to actuality.

From image to object, from idea to thought

The goal of a drive is to discharge its cathexes. A cathected idea can discharge in a (perceptual) surrogate such as dream or hallucination when an appropriate motor outflow is unavailable. But this is retrograde conduction: 'the only way in which we can describe what happens in hallucinatory dreams is by saying that the excitation moves in a *backward* direction. Instead of being transmitted towards the motor end of the apparatus it moves towards the *sensory* end and finally reaches the perceptual system' (SE 5: 542). The possibility of a reversal of psychic flow, in this example, or in others, such as the attention-cathexes, conflicts with microgenetic theory, which is anisotropic with respect to the direction of mental process.

In cases with brain damage, the banality of most hallucinations, most symptoms generally, makes it unlikely they are surrogate gratifications of drive or result from a concentration of cathexis on a memory trace when discharge is not possible. In such cases, hallucinations occur as incomplete objects, or as perceptions that do not fully exteriorize (Brown, 1988). They are brief intrusions in object perceptions. The continuum from image to object (see below)[11] and the relation of hallucinatory images to the perceptual process, e.g. hallucination replaces perception, utilizes common brain mechanisms, etc., argues against the psychoanalytic concept that hallucinations are repressed memories, or that perceptions are sensory systems without content. This was argued most forcefully in the *Project*. Later, Freud wrote, 'the *Pcpt.* system has no memory whatsoever' (SE 5: 539).

The sharp distinction of memory and perception was an unfortunate turn, for psychoanalytic theory, as for psychology in general. Dreams are hallucinations that spread across modalities to achieve an almost veridical clarity. Like myths, they have an historical character; they represent the individual's needs and strivings. So do perceptions, but they have adapted to the point that their drive origins are inapparent. Dreams are partly memories, but memories, like perceptions, do not call for interpretations. That is because memories occur in relation to waking awareness, closer to perceptual reality, while dreams, which are the sole world of the dreamer, are another reality, and thus in need of explanation if the objective world is to be conserved. In dream there are no comparisons. Memories feel like past events by a comparison to the present, while dreams, having only a present, have no past or future to compare with.

Instincts remain unconscious since a libidinal cathexis cannot cathect itself. However, ideas, or the ideational representatives of instinctual states, such as hallucinations, can pass from an unconscious state to consciousness. This could be taken to imply, accurately I believe, that only conceptual entities can become conscious or, put differently, that consciousness is fully conceptual, i.e. what enters consciousness is part of a concept or category, though feelings that become conscious are not, in the psychoanalytic model, conceptual entities. Feeling seems to take a different route. Freud says, while feelings can be transmitted to consciousness directly, ideas become conscious by way of associative links to words (e.g. SE 19: 22). If the drive energy ingredient in the idea is modified to a feeling that enters consciousness by virtue of this attachment, do ideas or concepts not shape those feelings that reach consciousness by a separate path? This means that every idea has an affective tonality, and every feeling an ideational content. Indeed, if feeling is a tributary of drive, what else could give the feeling its identity if not its conceptual or ideational content?[12]

The path from image to object is replicated in the development of memory to thought. Freud gave a description of 'a train of ideas' as follows: 'We believe that, starting from a purposive idea, a given amount of excitation, which we term "cathectic energy", is displaced along the associative paths selected by that purposive idea. A train of thought which is "neglected" is one which has *not received* this cathexis; a train of thought which is "suppressed" or "repudiated" is one from which this cathexis has been *withdrawn*. Under certain conditions a train of thought with a purposive cathexis is capable of attracting the attention of consciousness to itself . . .' (SE 5: 594). The purposive energy 'is diffused along all the associative paths that radiate from it.' In other words, a purposive idea excites relevant traces, neglects others and is impeded from exciting those traces from which excitation has been withdrawn, i.e. subjected to repression. The theory is a restatement in neurological terms of a folk-psychology of everyday experience. MacIntyre expressed the views of many commentators when he wrote that, 'Freud preserved the view of the mind as a piece of machinery and merely wrote up in psychological terms what had been originally intended as neurological theory.'

In Freud's view, the purposive train of thought is wholly rational, whereas the suppressed train is 'drawn into the unconscious'. A deviant thought is a branch of normal process, not as in microgenesis, an intrinsic phase in normal cognition. As early as the *Project*, Freud attempted to relate the psychoanalytic model to normal cognition, but the dual-route for neurotic and rational thinking (see below) tended to vitiate the interpretation of pathology as a model of the normal. Similarly, there were difficulties connecting the model to organic states, as is evident in the labored comparison of symptoms and errors (see below). In a momentary state, where pathology is not a definite factor, one would not want to say that all of the contents other than those that actualize are repressed or excluded by the defenses. Certainly, one would not expect different sources or types of cathexes, or different paths for normal and pathological cognition.

The principal mechanism by means of which an unconscious idea raised to the system preconscious achieves consciousness is by an association to verbal traces in the preconscious. A reduction in the anti-cathexis also favored entry to consciousness, by lowering the resistance in the associative path of the idea. Once the memory trace is activated, in order to reach consciousness it must be linked to a verbal image, i.e. the object concept must be named.[13] Freud wrote that the conscious idea comprises both 'the concrete idea plus the verbal idea corresponding to it whilst the conscious idea is that of the thing alone.' Or, with unconscious ideas, 'connecting-links must be forged before they can be brought into the *Cs*' (SE 14: 152). An unconscious idea anterior to words takes on quality through access to verbal memories when an object concept connects to a verbal image. The preverbal idea is perceptual, i.e. imaginal or hallucinatory. The connection to words involves the activation of word-memories and their influence on the quality of the visual idea. Freud's account of words as labels attached to objects was forecast in his book, *On Aphasia*.

In rational thinking, the wish is adapted to reality. According to psychoanalytic theory, this is not achieved by breaking the force of the drives, but by accessing memories independent of the libidinal path. Freud maintained that the memory trace could be the target of secondary process, and receive cathexes by another, purposeful route, though it is unclear, when a trace is revived through the ego cathexes of the secondary process, whether this process is voluntary and directed or still under the guidance, though muted, of the drives.[14] This description of two routes to the memory trace, one by way of drives, in the primary process, the other for rational thinking by way of the secondary process, adds new difficulties, including a reversal of the direction of the cognitive process and the role of conscious agency in the selection of unconscious traces. Simply put, how does *Cs* sort through the bramble of traces that are unknowable – as Freud claimed, noumenal – because they are unconscious? How does an agentive consciousness know what memory it is looking for before it finds it?

Guardians of consciousness

In general, Freud rejected the notion of altered states of consciousness in favor of mechanisms to explain how mental contents entered or were excluded from the system consciousness. The censorship that excludes unconscious ideas from consciousness acts on either boundary of the preconscious. How it works is uncertain. The location *between* systems is untenable once the concept of system-location is eliminated, i.e. when topography no longer matters, 'between' loses its meaning. The 'between-system' formulation has been described as a compromise activity employed by Freud 'to subsume distortion, censorship, and dream work' (G 100). The defenses, repression and the censorship, are distinguished partly on the basis of their outcome in compromise formation and in symptoms, and partly on the basis of the form of energy they utilize, but they appear to act through the similar mechanisms of withdrawal of cathexis and anti-cathexis. These mechanisms, taken as givens, provide an explanation upon which the hierarchy of defenses is layered.

The incomplete deactivation of competing ideas seeking access to the preconscious gives rise to doubt and degrees of uncertainty. Freud wrote that doubt is 'introduced by the work of the censorship between the *Ucs* and the *Pcs*' (SE 14: 186). On this model, doubt and its partial resolution in choice are products of unequal censorship. The censorship is described in terms of countercathexes that function as gates between compartments, or possibly *within* systems, filters perhaps, but requiring a decisional operation that is almost homuncular, not between options but as to the locus and relative intensity of their application. In fact, while not offering an explanation, Freud tacitly acknowledged the problem of explaining the selectivity of the censorship by cautioning his readers not to consider it like 'a stern little manikin' or a 'little chamber of the brain'.

In contrast, one could align doubt, uncertainty and choice on a microgenetic continuum of increasing definiteness. Choice is more concrete than doubt. It is a phase in the derivation where a final option has not fully individuated. The individual takes a step back from commitment. Unlike doubt, which is centered in the subjective, in one's strengths, weaknesses and capacities, choice tends to be object-centered, where capability is less the issue than the adaptiveness of competing outcomes. From the psychoanalytic perspective, doubt should point to the transition from the unconscious to the preconscious, while choice should point to the transition from the preconscious to consciousness. In microgenetic theory, a mechanism of censorship that introduces doubt is replaced by a segment in the becoming of a mental state that elaborates a private space of introspection in which doubt, uncertainty and choice arise as an accentuation of a *subjective* phase in the aim to objectivity.

Repression (of ideas, affects) is an inference about a mechanism (decathexis, countercathexis) that explains the non-occurrence of an anticipated or appropriate behavior by the appearance of a symptom. The linkage of the symptom and the intended, but repressed, target is explained by the energetics of the drives. Elsewhere, I have discussed a model of displacement that does not require cathectic dynamics. The model entails a specificity to content at each phase in its derivation. Thus, infantile ideas and fantasies in the unconscious are 'repressed' or 'barred' from consciousness by the organization of those ideas in primary process mentation, i.e. their 'state-specificity'. The unconscious content and its cognitive mode (style, strategy) pass by degrees to consciousness. Gill suggested that the lack of access to consciousness is state-specific, citing the disparity in organization between primary and secondary process mentation. Again, this disparity should be non-selective and involve all primary process content, since a content is not an isolate in its mode of mentation, whether its becoming-conscious is a continuous transition or a mapping from primary process fantasy to secondary process verbal memory.

Freud wrote, a symptom 'denotes the presence of some pathological process . . . when a function has undergone some unusual change or when a new phenomenon has arisen out of it' (SE 20: 87). Here, the symptom is not a phase in the normal but a byproduct. Leaving neural concepts aside he goes on to say, 'a symptom is a sign of a substitution for an instinctual satisfaction which has remained in abeyance, it is a consequence of the process of repression. Repression proceeds from the ego when the latter – it may be at the behest of the superego – refuses to associate itself with an instinctual cathexis which has been aroused in the id' (SE 20: 95). While analysts have found Freud's theory of the symptom useful for clinical treatment, there is little to recommend it on metapsychological grounds, nor does it speak to the problem of organic symptomatology. An aphasic error, an organic hallucination and a developmental anomaly of language or behaviour are not motivated surrogates arising from repression or censorship. Yet, if we do not relate the organic to the neurotic, we do not have a comprehensive theory of the symptom.

Gill writes, 'the defenses constitute a hierarchical series which regulate originally *ad hoc* discharges running from primary to secondary process and from drive to inhibited cathexes. The more primitive the place in this hierarchy occupied by a particular defense, the more closely is it connected with repressed material, the more likely is it that it regulates originally *ad hoc* primary-process discharge of uninhibited energy, and the less likely is it that it can become conscious' (G 115). The description of layers of defense, from the most primitive and unconscious to the most conscious and adaptive, has some points of contact with microgenesis, in that the conscious realization of an unconscious idea reflects a traversal over layers of constraint in which a sufficient portion of the potential of the unconscious content survives as a compromise between final deviance and adaptation. The layers of defense and the intensity of the censorship would then correspond with inhibitory sculpting in a process of actualization. The problem with the analogy is that, in psychoanalytic theory, the sculpting leads to distortion, while in microgenesis, the 'distortions' are eliminated by the sculpting.

The difference goes to the heart of process theory. In psychoanalysis, process is displaced from the center of the content, which is conceived as a logical solid, to its periphery, which is conceived as a surface for energetic exchange. The content is secondarily distorted by extrinsic drive, e.g. the wish that discharges in dream, or a substitute emerges as a symptom that is a more or less distant relation of the desired goal. In contrast, in microgenesis, the dynamic of quantitative energy is 'moved' from outside the idea to inside the content as qualitative change. There is no stable entity, the idea, that secondarily changes, rather, there is change that deposits a relatively stable idea.

Repression and symptom formation

Neuropsychological studies confirm that the mode of consciousness accompanying a pathological fragment, in perception, action or in language, is specific to the modality and the momentary symptom-form. The neuropsychological literature on this topic is impressive. We do not consider a retrograde amnesia a result of repression. Events in the right hemisphere of a callosal sectioned patient are not verbalized, but we would not say they are repressed. The failure to recall the experiences of early childhood, like the inability to recall most dreams, or 'right hemisphere' cognitions, or aphasic blocks, can be better explained by the dynamics of normal cognition, the microgenesis or actualization of mental contents out of antecedent potential, and the state-specificity of configural entities at different phases in this process.

With regard to language, an aphasic who cannot recall a word does not repress it, nor is a substituted word a symptom of repression. An individual may be aware of a correct word in an utterance but unaware of the incorrectness of another. The error (paraphasia) is generally related to the target by abstract word-category relations, for example, saying 'table' or 'throne' for the word *chair*. The more remote in meaning the error is from the target, the less the subject is aware that it is incorrect.[15] The remoteness of the symptom is also a factor in psychoanalytic interpretations. Freud wrote that *Pcs* cathexis 'attaches itself to a substitutive idea which, on the one hand, is connected by association with the rejected idea, and on the other, has escaped repression by reason of its remoteness from that idea' (SE 14: 182).

Paraphasia is a microcosm of the substitutive idea, or a substitute by displacement. A word such as 'chair' is a member of a category, such as furniture. Indeed, chair is a typical (prototypical, central or core) member of this category. Other members of the category, such as a lamp or a carpet, are less typical, while still others, such as a piano, a painting, or a fireplace, are extrinsic to the category, or have features that place them in other categories. A patient may name a chair as a table, where the substitution is close to the target, i.e. both items are typical members of the category. This is a common response. But the patient might also name a chair as, say, a 'window', where relations of proximity or experience are more apparent than those of abstract category. In this instance, the error violates the category boundary. What determines whether the misnaming will be close to the target or distant? I think the error type reflects the phase that is disrupted in a process of semantic realization. This process goes from wide to narrow targets, and gradually 'zeroes-in' on the correct word. The specification correlates with a genetic derivation over distributed neural systems.

How does this relate to the concept of repression? In Freud's account, the symptom is a defense against repression, i.e. a surrogate idea that arises from the inactivation of the painful idea. Freud writes, 'a symptom arises from an instinctual impulse which has been detrimentally affected by repression [SE 20: 94] . . . (to bind) energy that would otherwise be discharged in anxiety' (SE 20: 144). Repression is a way of explaining the connecting links from the repressed idea to the manifest symptom. Freud seems to distinguish symptoms from errors. In the *Project*, he wonders, 'how *error* can occur in the course of thought. What is error?' But the account is convoluted and inconclusive. In microgenetic theory, the nature of an error (symptom) reflects the proximity or the remoteness of the disrupted phase to the presumed target, i.e. it is a function of the 'locus' of disruption in the conceptual or semantic 'space' traversed by the developing configuration. Semantic distance is not related to the intensity of decathexis or the evasion of the idea that is repressed, it is a marker of the cognitive depth, thus, the affective strength, of the disturbance.

Moreover, an idea is not altered in a vacuum but develops in a setting of neighboring ideas and memories, along with other aspects of mentation, including affective tonality. As in the above example, consciousness is differently allocated to – generated by – the transforming segment with an attenuation of that portion mediating the derailed item. The symptom, the incorrect word or its preliminary meaning-content, is a piece of unconscious mentation embedded in a behavior that, apart from the error, is otherwise fully conscious. That a symptom as discrete as a word can accompany a lack of focal critical awareness is problematic for psychoanalytic theory. Why should a searchlight *Cs* ignore a semantic error, while it recognizes a phonological or grammatical one? There is no evidence that most organic symptoms are motivated, or reflect the personality of the individual, or are distortions of verbal memories resulting from anti-cathexes.[16] They are lawful, and tend to occur in predictable classes that are comparable across cases. The implication for the study of consciousness is that consciousness is not a system or function attached to a content, but a state that is realized through multiple contents at simultaneous points in the course of an object development.

From this point of view, repression can be reinterpreted as follows: either a global content with intense affect and reduced awareness enters consciousness, or a partial content with reduced affect and greater awareness. (In these and other examples, quantitative terms are shorthand; every change in quantity is really a qualitative adjustment.) In the first instance, the unconscious construct actualizes without a specification of its content or feeling and without a development of a critical awareness, i.e. consciousness and its content are global. Or, one could say, content and feeling fail to individuate awareness. All potential avenues of individuation of the global construct are latently possible and each, were it to develop, whether as a symptom or an adaptive behavior, would express some portion of the potential from which it develops.

Conflict and adaption

Rapaport wrote that repression begins with the withdrawal of the cathexes of the conscious and preconscious system from the idea to be repressed, followed by the countercathexes of these ideas to prevent their re-emergence. The anti-cathexes derive, paradoxically, from the very drive cathexes they counteract. Freud wrote, 'It is very possible that it is precisely the cathexis which is withdrawn from the idea that is used for anticathexis' (SE 14: 181). This would explain the selectivity of the anti-cathexis, i.e. its origin in cathexis fixes its locus in those targets, even if we are still lacking an account of the specificity of cathectic energy. The anti-cathexis derives from the preconscious registration, since ideas are active in the unconscious even when they do not reach consciousness. Anti-cathexis, like cathexis, is drive-based, originating in *Pcs* energy, perhaps in energy bound to *Pcs* ideas, or in affect-repositories which are the preconscious representatives of drive. The anti-cathexis cannot originate directly in drive energy, which is unconscious and contrary to its aims.

Evolutionary conflict tends to be located at the interface of the organism with its environment, though the struggle to survive goes on at every stage, from the genes to behavior. Psychoanalysis acknowledges the role of struggle and adaptation, yet the metapsychology is a theory of conflict displaced inward, first to the surface of a neuron, then to an idea. The conflict is not between competing ideas but competing wishes, or desires, the seeking of pleasure or the avoidance of pain, or the frustration of a wish, all deriving from the conflict of the excitatory and inhibitory energies of the drives acting on the same idea or its associates. (A fuller explanation of this point requires discussion of affect theory, which is beyond the scope of this chapter.) The activation or inhibition of an idea is determined by its compatibility with other ideas, or by the 'meaning-content' of a wish. A wish is for or about something, and that something entails a concept or an object, so that the energies impinging on ideas derive from other ideas by way of the energy attached to them, even if psychoanalysis attributes the source of this energy to the drives. Here, Freud must be amended if the theory is to be coherent.

On the other hand, microgenesis is a theory of cognitive adaptation in which contrast and constraint take the place of competing energies. The different affect-intensities of unconscious and conscious contents result from the individuation of drives to feelings, and conceptual primitives to partial concepts, as an act of cognition develops from an unconscious to a conscious phase. The differentation of an idea accompanies a mitigation and qualitative refinement of its affective tonality. A conscious thought may be associated with intense feeling, but a thought so charged differs from one in which the affective tonality is reduced, e.g. in its selectivity, intentionality, etc. The latter is more discriminative, options, including detachment or inaction, are possible; the former is impulsive, options are foreclosed and actions feel inwardly coerced. What survives of an affec-tively-charged idea in consciousness is not the original idea minus the drive energy, most of which has been trimmed away in its journey to consciousness, but what, given the resources of the individual, is most adaptive at every phase in its actualization.

Neuropsychological studies demonstrate that anxiety is a symptom of irresolution, when objects have not individuated to full clarity. This is not equivalent to conflict in the psychoanalytic sense. There is a substantial, though largely unfamiliar, literature on this topic dating from the percept-genetic research of Friedrich Sander, continued by Gudmund Smith and his colleagues in Lund, among others.[17] These studies employ techniques of rapid tachistoscopic projection of visual stimuli to elicit anxiety from object pictures that are incompletely analysed. The anxiety can occur with neutral stimuli, though emotionally-laden scenes may heighten the response. Clinically, the same phenomenon is observed in aphasics searching for the correct word, and in other brain-damaged cases. Goldstein's term, 'catastrophic reaction', aptly describes the anxiety and frustration that can occur. The anxiety points to a phase in the derivation of drive to feeling, or concept to object, where the preobject and its affec-tive charge have not yet resolved to awareness. Repression is not an issue. We are dealing with picture description, or the naming of a chair, not incestuous memories.

In psychoanalytic theory, the passage from an unconscious idea to a conscious thought involves a progressive bottom-up migration, or bottom-up and top-down energizing, of a content in the systems *Ucs* and *Pcs*. This process can result in a conscious adaptation but, being drive-based and originating in primary process, it is inevitably irrational. In contrast, logical thought has its origin in secondary process, which is able to cathect the trace into consciousness directly. Neurotic symptoms are a clue to the unconscious origin of the idea. They represent the arousal, in primary process, of primitive traces deflected by libidinal energy, in contrast to rational thought which represents the purposeful cathexis of unconscious ideas by secondary process. One would expect a continuum from the neurotic to the everyday but Freud distinguishes mechanisms of neurotic ideation from those of rational thought. Does this mean that any thought that is not a logical proposition is neurotically based?

Cathexis, decathexis and anti-cathexis are descriptions of the dialect between the latency, ecphoria and valencies of memory traces, the facilitation and de-facilitation of the access of memories to consciousness, or conscious access to memory. In both the topographic and dynamic models, the autonomy of the trace corresponds to the selectivity of the charge. A trace or idea marked off from its context is an entity that can relocate to consciousness or become conscious independent of its context. Indeed, the context imposed on the trace by drive is independent of the organizing activity of memory. The autonomy of the idea corresponds with the autonomy of the trace – the idea *is* the trace plus activation. The trace, in turn, is a copy of the original impression. The transition from perception to memory trace to idea involves the same content. The dynamic is removed from within the content, where it belongs, to the surface of the content as an extrinsic impulse which is then reintroduced to explain how the content undergoes transformation. The model, to its credit, does have some points of contact with current theories of memory, even if, in my view, those theories are equally deficient, where a content is looked up and retrieved from a storehouse as a copy in consciousness.

In sum, according to this theory, a trace awakens when it is activated. When the trace is not activated, it remains dormant. When it is activated, the activation may be countered by an opposing deactivation. Both activation and deactivation are drive-based and motivated. The activation can arise in primary or secondary process. The trace is independent of what activates it. A trace from which activation is withdrawn may discharge its energy in the unconscious or in consciousness, as neurosis, dream, etc. The trace can be further activated in (or transferred to) the preconscious, at which point, through association to verbal images, it can become conscious. The activation (cathexis) of a trace does not guarantee consciousness, but is the potential to become conscious if the content can attract an attention-cathexis or form associations to preconscious verbal images.

From unconscious to conscious mentation

The primary process mechanisms are displacement, condensation, substitution and symbol formation. These are the mechanisms of a primitive cognition. They act through the dreamwork and the countercathexes of the censorship to produce distortions of unconscious desires. A symptom is a sign that this process is unsuccessful. Psychoanalysis differs from microgenesis in that, in the former, ideas uncorrupted in the unconscious undergo distortion whereas, in microgenesis, unconscious contents begin as 'distortions' and develop to adaptive specificity. The idea does not undergo secondary distortion; primitive or primary process mentation is deviant, by definition, in comparison to waking thought. The distortion reflects an incomplete conceptual or semantic analysis, with derailments of content on the basis of perceptual and meaning-relations. The incomplete analysis manifests itself in a prelogical or syncretic mode of cognition. In fact, the urge to interpret a dream may well be a conscious continuation of the unfulfilled process through which dreams are normally 'analyzed' to waking content.

The mechanisms of the dreamwork reflect the unification by the mobile cathexes of widely scattered but related unconscious or preconscious ideas into a drive-based motivated whole. The cathexes and associations that underwrite these interactions give way to the arousal of ideas by a configuration actualizing through conceptual and semantic fields, out of which words are elicited from semantic and phonological contexts. Dream is the prototype, or the product, of unconscious cognition, and the comparison of dream to waking is a comparison of unconscious and conscious cognition.

Are we correct in bifurcating mentation into two (or, including the system *Pcs*, three) modes of ideation or thought? Perhaps there is a gradual derivation to consciousness, one in which dream does not play the part of a pathological side-branch but a transitional phase. Well before Freud, Herbart maintained there was 'a gradual continuity in the passage from actual ideas to the slumbering ideas of memory, and conversely, as well as the possibility of a reciprocal action of these slumbering ideas, without condescending to a materialistic mode of explanation of these processes, in the sense of seeing in them only material cerebral processes of a strength insufficient for incitation of consciousness' (PU 1: 35).

Rapaport, still influenced by the quantitative concepts of after-discharge, safety-valves, tension and discharge, attempted to do away with 'the arbitary segregation of conation, cognition and affection . . . of memory, association, imagination, etc.', by postulating that ideation 'changes into thought, which is experimental action with small cathectic quantities; (and) has available to it – ideally – all memory traces and their relations, for orientation in reality' (R 423). The distinction of ideation from thought, the former being the primitive in mentation, the latter, the rational, might be construed as indicating that thought develops out of a 'primitive' phase.[18] However, Freud argued that an act of cognition is not a passage from the primitive to the rational, though culture progresses in this direction, but rather, the primitive is an offshoot of the rational, not its immediate precursor, for unconscious ideas can be directly cathected to consciousness by secondary process.

Rapaport was closer to Schilder than Freud in this respect. Citing additional work by Kris and Hartmann from a microgenetic standpoint, he described a *continuum* from primary to secondary process, or from ideation to thought: 'I have observed a more or less continuous series of such states of consciousness which seemed to me distinguishable from each other by the formal characteristics of their mode of thinking. The differences in cathectic dynamics between these states of consciousness and the forms of the thought processes determined by them are so far unknown. A field of exploration lies wide open here, and a huge gap in our theory of thinking awaits bridging' (R 323). This continuum has been documented in the *organic* pathologies.

The notion of a continuous series in the derivation of a cognition is fundamental to microgenetic theory. The actualization of the mental state is conceived as a phase-transition from an unconscious core to the world surface. Freud was correct in postulating that memory is the beginning of thought, but he erred in conceiving perception as prior to memory, for then he could only interpret dream and hallucination as the (defective) products of the revived traces of sense-impressions, not as preliminaries to object perception.[19] A microtemporal account of the perceptual process has the 'memory trace' laid down at successive moments in the object-development, rather than being deposited after the object is perceived.[20]

On this view, the 'trace' is not re-represented or conveyed from one level in the mental state to the next but consists in the full sequence from onset to actuality. The progression from configurations in which the mnestic is more prominent than the perceptual, e.g long-term or personal memory, to those in which the perceptual and mnestic are compresent, e.g. short-term or iconic memory, to an external object that *appears* independent of, or prior to,[21] one's memory of it, is an essential feature of microgenesis. Specifically, for microgenesis, the conventional direction of memorization from perception to a long term store is reversed; remembering and perceiving are one process, in which anticipatory objects are apprehended as memories, and external objects are the final phase in revival. An object-perception is an actualized memory constrained by sensation to mirror the external world. Without such constraints, a perception does not achieve objectivity, and is thus consigned to a manifestation in memory, for example, as a dream or an hallucination. Every object has a mnestic, a perceptual and an affective aspect at every phase in its formation. Memories become facts as instinctual preobjects transform through conceptual feeling to external entities and their valuations.

The microgenetic interpretation of affect as intrinsic to ideas, not attached them, contrasts sharply with the psychoanalytic notion of drive attachment. We think of ideas and feelings, concepts and emotions, as separate components of mentation, and this tendency is accentuated by the comparison of animal and human cognition. One could ask, where are the concepts in animal rage, where is the emotion in human logic? An instinct shifts to a concrete object-concept when a goal becomes explicit. In experimental animals, brief latency undirected 'sham rage' points to a display mechanism without inner feeling, a brainstem device at the periphery of cognition. Yet, normal rage in an animal is directed to a specific object. For example, anger is directed to its provocation. The isolation of a target betrays the concept in the drive. The object is the key to the category of the drive. Every concept entails a feeling, every feeling assumes a concept.

If the energy of drive inheres in the idea, what is the nature of this *inherence*? Presumably, one or a few primitive concepts and their drive content, unified at the start, initiate the cascade of affect-ideas and object-desires. The drive or affect-content of a primitive spatial construct and its ensuing satisfaction, all part of the same configuration, press to actuality. The construct enfolds the drive and forcasts but does not yet contain its future object. The transition from trace to idea in psychoanalytic theory might be taken to correspond to that, in microgenesis, from a spatial construct to a phase of meaning-relations and personal memory. The delay before discharge that, for Freud, explained thought, is the embedding of whole/part shifts that 'stretch' the transition, permitting the drive with its idea to fractionate from a phase of spatial relations to one of conceptual feeling.

With further postponement of discharge, i.e. extending the microtransition, conceptual feeling parses to the sub-categories of anticipatory objects and their companion affects or affect-ideas, partial concepts charged with affect and meaning. The association of the idea to the verbal image in psychoanalysis corresponds with the transition from pre-lexical concepts to lexical items through fields of word-meaning relationships. In this process, drive becomes desire, and desire becomes a partial affect such as preference. A wish is an intentional drive, a drive that narrows down to an object or object category. A preference points to a choice within the category. Global acts and objects become intentions as they move outward from the body. Options emerge prior to final selection. Intentionality is a sign that an object (concept, image) has individuated, in the world or the mind. Volition or agency is a sign of access to alternative paths, latent in the potential, before a final object has actualized. With an active potential in the background, and a concrete world already realized, an expansion of that segment between the core of cognition and its world surface permits options and choices, enlarges the interior space of introspection and accentuates the feeling of agency.

Chapter 6
Consciousness and the Categories of Nature

The Lord: How many *Tathāgatas* have you honoured, Mañjuśrī?
Mañjuśrī: As many as there are the mental actions which have been stopped in an illusory man.[1]

So Prince Mañjuśrī replies, on his way to Buddhahood. A *Tathāgata*, a 'thus-gone-one', is not a subject or a nameable, but is to be experienced in the mode of suchness (*tathata*), a field of no relations, spaceless, timeless, a field of no-mind, no-thing, no-change. An action, a motion, a starting up, is an illusion, as is a stopping and a ceasing. So it is for the *Tathāgata*. The dharma, the law or process, is prior even to the Buddha. For the Buddhists, the contents and forms of sentience are the illusory guardians of the real; mind is a barrier to an experience of physical reality. The prism of distortion of the mental is lifted in enlightenment, a lifting that is a *seeing through*, as the lifting of a dream opens a door to wakefulness, so consciousness and its categories dissolve on awakening to the emptiness of all actualities.

Buddhist belief and process thought

From the *Yogācāra* of Vasubandhu to the idealism of Bradley,[2] the argument has been made that consciousness is the essence or mind-stuff of the universe, or that it is identical with the absolute or, less strongly, that consciousness, though a phenomenal simulacrum, is the only means by which the absolute can be approached even if, for the Buddhists, consciousness and its props – reason, desire, agency – dissolve into passing states of sensory experience, themselves composites of finer entities, fictions of the understanding, all the way down to the vibratory point-instants of absolute reality.

In Buddhist thought, a nesting of categories is thematic from one layer to the next over the entire system of thought. One might suppose that a theory on the composition of the phenomenal would spread to the categorical nature of the absolute, for example, that the mind-only school of *Yogācāra*, and the non-discriminative reality of *Mādhyamika*, would merge in an approach to consciousness as a starting point in a descent of categories from linguistic structures in the mind to the core of physical reality. This would reinforce the possibility that the categories are not illusory but foundational and that the pattern of categorical descent might provide a path from the screen of cognition to the bottom-most flickerings on which the entire world of illusion is elaborated.

However, for the Buddhists, a concept – any concept – is a phenomenal construction of physical instants and an impediment to a knowledge of the real. Through meditation and right action, the conceptual is stripped away from mind and nature, and with it the pattern by which the entire artifice of the phenomenal is assembled. With a dissolution of the conceptual, so it is maintained, there is surrender to the emptiness of entities as pure, tenseless, termless relations in a plane of non-duality beyond the shadow world of life's phenomena. This is an idealism that holds as counterfeit the subjectivism of its own base of knowledge.

The belief that categories are not a guide to foundational laws but assertions that contain the seeds of their own negation inheres in the dialectical framework of Buddhist logic. A category is a boundary, a demarcation or a contrast, and a boundary is a line around which an opposition can form. The fact of a thing supposes its factual non-existence as an antithesis. An assertion is a perspective. Since the absolute is non-perspectival, the path to awakening tends to be described in terms of negatives. One thinks of the insubstantiality of becoming, the middle course between eternalism and annihilation, the divestment of strivings, the shedding of the self, of thought and deed, the futility of affirmation or denial, in sum, a disavowal of the conditions for the realization of all actualities.[3]

The response to the real in Buddhism is a retreat from the actual to an embrace of the indefinite, since the actual becomes definite when it becomes real. For the Buddhists, antecedent potential is coordinate with ultimate reality. Suzuki writes that emptiness is a zero full of infinite possibilities. This is a potential without content, a not-nothing that is not yet a something. Emptiness is not absence. In Buddhist practise, the arising of entities is progressively attenuated, finally cancelled at its inception in a contentless possibility. This gives the state of the 'non-arising of a dharma', an absolute everywhere, nowhere, eternal in its timelessness, infinite in its spacelessness. Mañjuśrī asks, 'has the *Tathāgata* realized that all dharmas are the same as empty space?'[4]

Categories and relations

Let us ask more deeply what is the nature of a category. In Buddhist thought, the non-substantial is equivalent to the mutually-dependent or the relational. A category is a set of dependent relations, but in what sense are relations categorical? The 'chunking' of parts into wholes, or the forming of categories, is usually conceived of as an activity of the mind, an activity in which the mind constructs a category out of relational parts. But the 'construction' of wholes from parts, like the division of wholes into elements, would seem to presume a category in the background. Parts cannot be simply summed to wholes, because without wholes to begin with there is no rationale for the integrity of the sum, that is, there is no procedure to explain how the units are correlated. The category must be there in the first place, in fact, a 'construction' must begin as a unity that is parsed and enlarged *from within*.

This supposes there are no 'free standing' parts. The presumption of a category *behind* the parts, which is the view that a part is always part of a larger part, is the first step in relational thinking. However, there is a tension between the relational and the categorical, even if both are inherent to the same mode of thought. The relatedness of an entity is its lack of limitation. A category is a limitation around a set of relations. A relation points beyond itself, a category entails a demarcation. The problem is to resolve relatedness with categorization, or to combine transitivity with boundedness. Relatedness is the negation of substantiality but categories transform relations into stabilities. Ultimately, this is the problem of how change creates objects.

A descent into a categorical system is a decomposition. When the system is conceived as an assemblage of parts that are put together like a chair, and the parts are conceived as the final part-objects that compose the complete structure, such as the arms of the chair, the back and the legs, the decomposition is a disassemblage that does not shred the bonds between the parts, since these are no more than additions from which the whole can be reconstructed. This is true as well in a conceptual deconstruction when the parts are conceived as subcategories, for there is no actual substrate that is decomposing. The analysis is an intellectual act that uncovers or discloses the constituent categories. The parts that are exposed in the analysis are still parts, not formative phases. In this way, the metanalysis of an object of thought can be distinguished from the analytic process of nature.

From entities to objects

What is the nature of this distinction? Is the process of object perception a mode of existence distinct from that of material objects, or do the latter exist, independent of perception, in the same manner as they exist in the observer's mind? In other words, is the act of perceiving the same kind of act as the 'act' of existing? My response is that perception is not a veil to nature but an exemplification of its laws. The embodiment of material objects in perception occurs through a process similar to that which deposits them in nature. There is a continuum from the actualization of a non-cognitive entity, to the actualization of a perception, to the inventions of conscious thought, which are metaphorical extensions, or discoveries, of a categorical process that is basic to material entities. *Perception points in two directions, to a non-cognitive nature in which it is grounded, and to conceptual thought for which it is a ground.* The categories of human conceptuality are signs of nature's own creativity. There is no cerebral Rubicon from mind to causal nature.

The category of a non-cognitive object is the duration of one full cycle of its phases, a segment of time carved out of the eternity of past and future time. In this cycle, an epoch of time emerges out of timelessness to be absorbed in the next epoch. The duration creates a minimal entity of a set of phases that are simultaneous until the final phase is realized, at which point the entity becomes a stability, and the simultaneity becomes an epoch of time. The significance of internal relatedness is that the duration over which the entity actualizes is its actual or incipient subjectivity. The subjective arises in the temporality of nuclear categories and spreads to abstract relations in spatial categories. Duration extends the subjective aim of natural categorization to its limit.

How is the procession of mute facts attributed to causal nature 'chunked' into non-cognitive categories? Such categorization is possible only if the inexorable order of the world, which is inferred from the succession of objects in perception, is conceived as the busy surface of a process that is, for the most part, inapparent or concealed. The concealment of the ubiquitous is a validation of its centrality. The necessities of thought are habitual and invariant, and so remain in the background.[5] The essential is inapparent precisely because it is invariant. In contrast to the multiplicity and variety of finite actualities, the change or becoming through which they develop is uniform. The richness of the world and the progression from one physical state to the next that pass for the content and pattern of nature's advance are, instead, its superficial marks, the inheritance of unfelt pulsations distributing into experience.

From object to class

Categories that combine multiple objects differ from those that generate them. For the intellect, a category of objects is a complex that is synchronous with its parts. A category of *chairs* is an abstraction out-of-time, analysed *as if* it were a structure of simultaneous elements, spatial and durationless. There is no development from category to member. The relations are abstract, ordered, formal. A taxonomy of chairs does not recur in the mind as a system through which the object is evoked.

In contrast, perceptions are sequences that evolve through an implicit temporal process to deposit as spatial wholes. The analysis is diachronic, the parts individuate. The perception of a chair is a graded series of phases. The chair consists of these phases, not the parts that emerge in the final percept. This is true for auditory and other objects as well as visual perceptions. The temporal structure of an auditory percept seems closer to the reality of a chair as a relational existent than does a visual percept. Yet, auditory percepts can be thought of as whole segments of sound, not concatenations.

A category of like objects rests on the relations among its members. The category of a single object consists of the internal relations that make it what it is. The former is to the family resemblance of a word or an object as the latter is to the process of its production, or as causal interaction is to causal persistence. Yet, the temporal category of a single object is the basis for the spatial category of similar objects. The rudimentary evolves to the complex by an iteration of natural laws. An object class is a relational set of meanings, an object, a set of transformational relations. As we move from concepts, to percepts, to the interior of the object, the object itself becomes a category, its duration, the phases in its objectification, its persistence, the minimal novelty in a series of recurrences, its change, the phase-transition embedded in its appearance.

Conceptual parts and categorical wholes

A concept requires, minimally, an averaging of past encounters, extracting what is common to all encounters as its core and retaining the variation across encounters as its scope. The concept of a chair permits a range of chair-like objects in the category, not just a paradigmatic chair that is selected from a diverse sampling of exemplars. The legs of a chair can be replaced by a pedestal, the back can be lowered to near invisibility, the seat can be widened, material, color, shape can be altered. At some point the chair transforms to a stool, a side table or a bench. The potential to be a chair, as in a folding chair, is sufficient for the object to be classified as a chair. Concepts and categories are supple, mutable; they can be created as needed.

A concept is a compass in a sea of categories, a category is an anchor in the flux of transition. The process of categorization invades its own categories to carve out object concepts. The parts of the object assume the status of members to the concept, comparable to the way that objects serve as members to the category. In categories, the whole determines the parts. In concepts, the parts define the whole. In the growth of abstract concepts, there is an overlap in the properties of diverse objects, such that the property itself becomes the category on which membership turns. Commonly, perceptual relations give way to functional relations as the category becomes less restrictive. The object is then less a member than an illustration. For example, the concept of a chair is conditioned by a set of relevant perceptual features. The concept of furniture expands a functional property common to the class, ignoring the perceptual attributes that define a specific object. In the category of furniture, an extrinsic property, such as the movable comforts of a home, becomes a whole that defines the objects as parts. Objects may have nothing in common except the property that serves as a grouping strategy, such as 'things to take on a trip to France'.

A concrete or empirical concept is a category in which the parts, no longer constitutive but definitional, serve as miniature wholes, not ingredient in a particular object but shared by similar objects. Properties are the bridges we exploit to unify multiple appearances, e.g. in a chair, the seat, or the use for sitting. In a concept, the criteria for membership replace an enumeration of members based on those criteria. The criteria for inclusion in the category of chairs come to define the concept of a chair, while the concept of a chair determines which items belong in the category. A relation of part and whole is essential in concepts, as in categories, even for the most general of categories, minimally, as a contrast of the category with the whole of its negation, or the sense of a still deeper unity that underlies all oppositions.

Abstract concepts and empirical facts

For the Buddha, such object concepts as gave rise to desire and confusion were the concrete particulars that emerged from the process of conceptualization. The closer to sensory experience, the more corruptible the concept. But abstract concepts were still interpreted as a source of error. Vasubandhu pointed to the lack of clear demarcation between general and empirical concepts, and argued for a continuity from the abstract to the concrete. The continuity, in fact, extends to the sense-data on which the entire fabric of the conceptual is grounded. Objects are constructions of sense-data continuous with abstract concepts.

In Buddhism, the object appearance is an inherent feature of the awareness event. A blue ball is internal to the perception. The awareness of the ball and its blueness and the blue ball as an object – the form that apprehends and the form that is apprehended – are part of the same cognitive state. The continuity of thought and sense-data is similar to that proposed in microgenesis, in which an act of cognition is a sheet of mentation from core to world surface. However, in microgenesis, a sensory quality such as a tone or patch of color is conceived as a *terminus* of perception not a starting point. It is an outcome of an analytic process through which concrete facts are specified out of antecedent context and meaning.

If 'sense-data' are cognitive, and if mental experience develops from sensible qualities to complex entities, it is natural to think that abstract categories develop out of concrete ones. For Buddhism, concepts are synthetic and combinatory; they result from a compounding of phenomenal elements to complex wholes. The sensible qualities are the sources of object-appearances, which in turn are the sources of abstract concepts or affect-ideas. In microgenesis, the object is conceived as the final derivation of an abstract concept that is seeking actuality in concrete life.

How is one to decide if sense-data are the proximate ingredients of object concepts or the distal outcomes of conceptual primitives, that is, whether cognition runs from outside-in, or the reverse? A child's first words, such as mama or dada, refer to a class of objects and gradually zero-in on the correct target. The word is less a label than a category, a proper name treated as the stem of a common noun that incorporates referents on the basis of similar appearance, not shared functional properties. Over-inclusiveness or wide semantic reference is a form of primitive abstraction. In maturation there is a parallel growth of abstract and concrete entities, the latter issuing from the former.[6]

Concepts are the configural underpinnings of percepts, without which the impingements of sense-data would be random shocks. Sense-data are conscious qualities that must be configured before one is conscious of them, for the associative links between the qualities do not account for the coherence of the percepts that are their presumed resultants. Whitehead wrote, 'science conceived as resting on mere sense-perception, with no other source of observation, is bankrupt, so far as concerns its claim to self-sufficiency.'[7] The response to this criticism in Buddhism is that the cement of illusion connects the sense-data to a fanciful whole.[8] This is in line with Hume's postulate that the mind contributes the power or necessity in causation. But the need for a 'psychic addition' to sensory primitives creates an interface between the mental and the physical that is more jarring than the insufficiencies of either extreme. If the 'connectivity' in the sense-data that accounts for the coherence in conceptual elaboration is not in the data, nor a subjectivity imposed on raw but conscious 'sensation', the coherence must indicate that the data themselves are generated out of background conceptual forms.

Conceptual feeling

A realization of the conceptuality or relationality of all modes of thought leads to an apprehension of the phenomenal basis of consciousness and a divestment of the illusory concepts that populate the phenomenal domain. In daily practise, however, what matters is not whether concepts are concrete or abstract – all concepts obscure the path to enlightenment – but rather the feelings they evoke or express. Even if concepts could be eliminated, or conceived as empty of substance or meaning, what comes of the feelings that animate them? What is the nature of conceptual life devoid of feeling? The category of furniture seems to be independent of a desire for a beautiful chair, but in everyday life there are no affect-free concepts. Even if one is indifferent to a concept, say the concept of a unicellular organism, the traces of feeling are present in the choices that are made, the focus of interest, the way the object is defined, as well as its value or realness.

A desire can be mitigated by the realization that the self, its perceptions and mental contents are phenomenal constructs, but can it be abolished without changing its conceptual structure? Concepts and desires are interdependent. A desire 'attached' to a concept is an empty wish for an illusory object. A desire is the confusion that arises from the false substantiality of a goal. The confusion of concepts is the demarcation of the flux of reality into object-appearances. The emotional tonality of concepts is conditioned by the individuation of the conceptual life. When someone says, 'I love that chair', or dress, or dog, or person, feeling is allocated with the concept in a gradation of intimacy that, in this example, tends to reflect the privileged status of animate entities. The allocation may vary, but the process is always the same. The distribution of feeling is linked to the partition of concepts or the specification of conceptual feeling that grows out of individual character.

We can imagine a concept without a feeling, for concepts, like experiential memories, can be revived with little trace of the feelings that accompanied them, in their encounter or discovery, but can we imagine a feeling that is independent of conceptual form? A feeling without an object, like an intentionless concept, is a pure affect without direction or form, a subjectivity empty of content. This is an emptiness not unlike that of Buddhist practise achieved through the elimination of grasping. The withdrawal of feeling saps the object of the value that accompanied it in the course of its objectification.

A feeling that is suppressed or withdrawn at one locus tends to be heightened at another. A feeling that is not derived outward with an object will recede to the intensity of a drive and permeate internal states of irresolution. The erosion of intentional feeling with indifference to objects – including the self and inner states – leads to the replacement of object-bound feelings with intentionless moods. Feeling is the value in meaning that gives signification to abstract relations. The eradication of value independent of content turns concepts into the phantoms of the objects they once designated. Conversely, feelings need concepts to support them. Concepts are vehicles for feelings that are otherwise directionless. An object is a concept made concrete with feeling and etched into consciousness by drive and habit.

Categories and reality

The degree of concreteness in the entity is usually considered to be analogous to its reality. Conversely, the greater the distance of an entity from the concrete, the more emphatic its insubstantiality. In Buddhist philosophy, and in process thought, concepts are no more phenomenal than the 'sensory qualities' on which they depend. The real does not rest on a comparison of idea and physical object, or abstract and empirical concept. Ideas and sense-data are the polarities of the mental state, not images of increasing distance from an external object, and thus progressively less real. If to be real means to exist, what exists is always real at every point in its manifestation. For many, the progression from conception to perception to intrinsic being or existence, or the spectrum from mental representation to physical reality, refers to levels of greater or lesser proximity to a standard of the real that supposes a rock-bottom world of atomic solids in relation to which the mind is a mirror of fading approximations. If the real is the external, the contact with perception is the zone of its vulnerability. If the real is what exists, there are no boundaries. This is the standpoint of process theory.

Thought arises in categories not bound to concrete perceptions and the immediate pressures of survival. This does not mean that abstract concepts are secondary elaborations. A core of categorical primitives may well be enlarged to the abstract and expanded when the concrete is suspended. It is to be decided whether categories are enlarged by the addition of new members or expanded from within to accommodate them. If objects are not the forerunners of general concepts but their realizations, as I believe is the case, the elimination of object concepts that is prescribed in the approach to enlightenment – the retreat from the particular to the universal and the surrender of all categories – is an attenuation, an *évanouissement*, of the terminal specifics of conceptualization and, to use Heidegger's term, an *unconcealment* of the abstract universals that underlie the individuation of concrete forms. Bringing such elements to consciousness is a way of making them concrete. A retreat from the conscious particulars of an experience, or an abandonment of all object concepts in a descent from the particulars to the submerged generic forms, is a falling-into the universal ungroundedness of all experience.

The elimination of concepts begins with an erasure of the stock of naive beliefs defined by external relations, then a surrender of object categories differentiated by relations among their elements. A realization of the artificiality of everyday boundedness liberates the individual from a misconception of the real. The central insight of awakening refers to the nesting of categories from consciousness down to the ultimate flashings, point-instants or *kṣaṇa* that are the 'atomic bits' of spatiotemporal process. The absolute is not uncovered by an elimination of the categories for all entities are ultimately categorical. The categorical nature of thought and perception is not an illusory screen to a pure relationality, but a guide to a principle of partition of non-cognitive nature. An awareness of the relationality of categories and the categorical underpinnings of relationality, and a recognition that categorization is the substantial in process even if all categories are insubstantial, is the doorway to a perception of categories as the substantial aspect of all relations. The categorical aspect of relationality is an explanation of how entities can be non-substantial, yet stable. Their conceptual or categorical nature is obligatory if a theory of the absolute as sheer transition is not to leave the phenomenal unexplained.

Whether the universe is conceived as animate or material, where the boundaries of the conceptual are unsettled the categories of existence are ways of specifying the limits of value when the absolute is probed by theory. An account of the conceptual implies a theory of the material. The mental and the physical are dependent concepts, not opposing doctrines. The one supposes the other to which it is a response. The limits of a theory of the physical, thus the boundaries of the mental, are exposed when anything in either category is described, for the description of an entity is a piece of the theory that supports it. For every category, there is another category just beyond its contours. In fact, every statement plumbs the depths of the presuppositions on which everything depends.

Consciousness and categories

If the stage a science has reached is determined by the degree to which it is able to question its basic concepts, the most basic concept is that of conceptuality itself. By conceptual I mean consciousness, semantics, whole/part relations and the categorical nature of mental events. Consciousness is a mode of categorization at a high degree of refinement. Conceptuality is the process of categorization in relation to objects or events. The categorization that underlies conceptuality is usually taken as given in the account of concepts as the criterion of a rational cognition. McDowell writes that it is essential to conceptual capacities 'that they can be exploited in active thinking, thinking that is open to reflection about its own rational credentials.'[9] This is a standard description that conflates a number of mutually-supportive terms that need to be unpacked and reformulated. What is meant by capacity, activity, exploitation, rationality, credentials, reflection? The aim, of course, is to distinguish critical judgment from non-cognitive operations such as perception or information processing, while my goal, in contrast, is to establish a gradation of conceptuality from analytic thought and introspection to the most basic non-cognitive entities.

What I have in mind with regard to the categorical nature of consciousness is reminiscent of what the Buddhists say about an object such as the body, that it is nothing more than its constituents, the head, arms, legs, organs, and so on. There is no entity *body* in addition to these components. The components also decompose, for example, the head contains the eyes, but what are the eyes other than a collection of smaller organs, all the way down to the ultimate bits of matter. The decomposition is not necessarily explanatory. A description of DNA at the atomic level would not capture the essentials of genetic coding. Each level deserves an autonomous description, but at least we understand that we are talking of concepts or categories, not actual entities. Categorization is the core of conceptuality and the foundation of consciousness and subjectivity.

Reality: virtual and otherwise

An object before us in visual perception is an image that arises from binocular disparity, a fused image that is uncoupled from the inferred noumenal object that evoked it. This image is virtual in that it is 'computed' from the disparity in monocular vision. The perceived object resolves the competing images on each retina, but does not reflect either image directly. The location, depth and perspective of monocular sensation are blended in the binocular image. Moreover, the image is not contemporaneous with an object in nature. The temporal lag in perception obviates the possibility of a direct perception of the real. In sum, the image is a virtual construct of competing inputs, and is generated subsequent to the reception of the physical data that constrain it to model an object.

In an analogous manner, a virtual image of a duration is created from a disparity between the surface of the present – the terminus of the actualization – and the decay point or floor of the immediate past.[10] The duration of the present is uncoupled from the passage of nature as an island of time that is felt to be situated between the past and future, or conceived as a reactualization of the past plus its change in the revival, but not as a continuous line extending from one instant to the next.[11] Thus, as binocular disparity creates a virtual image of object space, a disparity over phases in the actualization of the mental state creates a virtual image of temporal duration. These images are comparable in that they are derived from spatial and temporal disparities. The images are virtual because they map to brain activity, not their referents, because they do not correlate exactly with external space-time, and because they achieve stability by replacing the real change – the actualization process – within the object, or instant, for the fictitious change across them. The object achieves this stability by eliminating the process at its core, a dynamic that is not abolished but persists, unnoticed, and finds expression elsewhere, presumably in the power or necessity of causal efficacy.

Whatever actualizes, actualizes concretely. The concreteness depends on how object-like the actuality is, and the specificity of reference, not on whether the object is situated in the mind or in the world. The concrete has the appearance of persistance because the more object-like the image, the fewer opportunities there are for deviation in its recurrence. The concrescence of Whitehead, in which the abstract is the enduring, the concrete, the perishing, is one way of characterizing the process through which abstract forms objectify. Everything is appearing and passing away and being replaced by another appearing. The evanescence of an object is the other side of its persistence. The persistence is the sameness achieved across a series of revisions. Stability is an illusion created by a similarity of replacements. Every entity perishes, so its stability is determined by the degree of freedom in the ensuing actualization. The object must perish to be replaced. The same process that is responsible for the stability, i.e. the reinstatement of a near replicate, is the source of the perishing.

The replacement of self and world is apprehended as a causal succession. The self perceives its own productions – images, acts and objects – as momentary and successive. It does not apprehend the generation of an object, or the exchange of one perception by its successor. A succession of perceptual worlds is guaranteed by the gaplessness of 'joints' between replacements. The shift from the duration of a self and world to the category of a self over time, i.e. the identity of the self, or a category of like objects that gives the similarity of objects in a class, is a shift from the natural category of becoming to the categorical artifacts of language and metacognition that propagate the virtual fields that becoming creates.

Whitehead argued that the awareness of reality is stratified into *simultaneous layers* of becoming whose temporal thickness depends on the specious present. A certain thickness of layering is necessary to permit the disparity across those layers from which the present is derived. The layering of the specious present corresponds with the thickness of the present that is realized. The thickness punctuates the duration, or reflects its punctuation by the stratification, and so stretches its subjective length. The layering has an elastic quality; the now is not a fixed duration. Repetitiousness, novelty, expectation, boredom, drowsiness, fever, sleep, all influence the duration.

On the possibility of pure consciousness

The goal of yogic practise is an attainment of a state of pure consciousness through a systematic elimination of the duality of subject and object. Meditation proceeds from the suspension of reflective and discursive thought to the elimination of discriminative perception and mental content. The initial reflection on an object is gradually revealed as an examination of the conceptual products of one's own imagination, actually, a state of reflection in which those products are no longer *uncorrupted*. The conceptual roots of the object-experience grow less transparent as the subject apprehends that every realization is developing over the path the prior object laid down, so that in complete absorption, gazing deeply at the inner object, its surface begins to fade by virtue of a sustained replication of its outward form. In this way, pre-attentive phases come to fill the conscious field with the potential in the prior state of what was left behind in the process of objectification.

It is questionable whether the endpoint of meditation in a state of pure consciousness is a lucid awareness devoid of constituents or a non-cognitive state of mere sentience. A state of pure subjectivity without a world, or a state of pure objectivity without a self, are the interchangeable goals of soteriologic meditation. In such a state, the Buddha realized the non-substantiality of all phenomena, but does not the realization of an idea require a subject who apprehends a mental object? The Buddha concluded that the awareness requires the availability of the sensory faculties and the bodily, verbal and mental dispositions.[12]

Whether in mystical union with God, the selflessness of samadhi trance, the unmediated awareness of animal perception (is awareness *mediated*?), the object and activity awareness described by Piaget in young children, or the *durée pure* of Bergson, many of us would concede the possibility of a non-dual consciousness. One aspect of the non-dual is the unity of oppositions. The apprehension that the perceptual world is an extension of the mind, or that the mind is continuous with material nature, dissolves the self-world boundary and, with it, all internal and external objects. How are such states to be interpreted? If the feeling of the duration of the present could be retained in the absence of a distinction of self and object, would the self not be eliminated, and with it the awareness of the state?

The category of consciousness rises and falls with its contents, but is not generated by them. Subject and object are enclosed by a duration. If acts and objects in the category of consciousness are construed as instants in the category of duration, like instants could they dissolve into the duration they articulate? If so, is a state of pure consciousness the apprehension of a duration expanded in time and emptied of momentary objects? In some sense, this is what happens when contents recede into the background on falling asleep or in snow-blindness, but in such states a normal consciousness is replaced by sleep, dream or confusion. When the object field undergoes diffusion, the self rises to prominence. When the field undergoes analysis, the self is diminished. We can attend to the components of self and object, but rarely to duration. Does a pure object-less state of consciousness entail a shift to this quieter aspect of the conscious state?

In brief, can we give up all objects and still retain a conscious duration? A loss of objects might leave the self the only remaining object, an intrapsychic object over the entire object field. The attenuation of the becoming would foreclose a present and its duration. The state would be objectless, presentless and durationless. What sort of consciousness would remain under these conditions? The change from one actualization to the next, and the stacking of events in the mind, provide the duration with a greater or lesser interval, the 'length' of which is determined by that event which, for the moment, comes to form the boundary of the immediate past. It seems certain that the posterior limit of the present can vary, thus the experiential duration, but events are still necessary to articulate the duration and a self is necessary to experience it. The self needs its objects but can give up its *attachments*. However, to attribute a pure consciousness to a duration without object attachments leaves unresolved the question of a duration independent of objects.

The antecedent limit of the now is established by the revival (decay) of a prior now in the occurrent state. A conscious duration of a second or more is the felt 'distance' to this phase extrapolated as a line to the immediate past. The posterior limit of the now is determined by the revival. A variation in this limit, in states of intoxication or fever, possibly in meditative states with a heightened sense of object or image awareness or of the flux of thought and the transition of mental states, might cause the now to dilate or contract. In confusion, dementia or amnesia the attentional span is reduced. There is distraction with shifting of interest and gaze, and an inability to maintain a focus on environmental objects or endogenous contents. The subject and his world become unstable. Might these conditions be explained by a present that *shrinks* in duration? If so, is it possible that the present might *expand*?

Consciousness and duration

For Whitehead, the consciousness of nature as a process did not imply that consciousness is a process. Consciousness is experienced in, and through, a present that is off-line with material process. He writes, 'the moment of consciousness involves a specious present in which there are antecedents and consequents'. To be a process in step with nature, like nature, consciousness would have to put the past behind it. In the becoming of non-cognitive entities, i.e. in the passage of material events, the revived past creates an implicit, non-temporal duration over its transitional phases. In the generation of a conscious state, the disparity in these phases is the basis of the illusory duration of the now. In the former, the antecedent and consequent are ingredient, in the latter, they are temporal limits in the subjective immediacy of the now.

Whitehead goes on to say, 'by an indefinite enlargement of the specious present, we can imagine an awareness of all nature as a process, although no process is implicated in the mode of awareness'. Whitehead is arguing that in the idea of an extension of consciousness or the specious present over all nature, the consciousness of all nature becomes the consciousness of God. Whitehead's God, like consciousness, has an antecedent and consequent nature. God is incomplete, like nature, a process of becoming.

The conception of a deity, or an ideal sphere that transcends physical law, such as the infinite duration of the pure land or Buddha field, is derived from an imaginative enlargement of a present moment uncoupled from physical process.[13] The present hovers over the passage of physical instants as the consciousness of God hovers over an eternity of passage. Or, given the idea of an infinitely expanded present in which all process is subsumed, an individual consciousness can be derived as a contraction of a God's eye present to the duration of a momentary human perspective.

This idea has appeared in other contexts, Augustine's, 'this thy today is eternity', or Boëthius, 'just as you can see things in this your temporal present, so God sees all things in His eternal present' or, more impersonally, preferable, perhaps, from a philosophical standpoint because it lacks the intimacy of ecstatic union, 'the passing Now makes time, the standing Now makes eternity'. Meister Eckhart wrote that 'the now-moment in which God made the first man, and the now-moment in which the last man will disappear, and the now-moment in which I am speaking, are all one in God, in whom there is only one now.' In Hinduism, the duration of the universe from its beginning to its end is one day, one kalpa, a single dream, for Brahma.[14] In meditation, 'a contemplation on time (is) directed towards the immediate realization of ever greater and greater durations and pursued until the whole of time can be experienced now.'[15]

The creation of the world out of the illusion of time, as in the 'eternal dream time' of the Aborigine, or the *Yogācāra* doctrine of the birth of consciousness and objects each moment from *ālaya-vijñāna*,[16] the bed of the unconscious, is a description of the punctuation of the timelessness of pure duration into the temporal epochs of mundane experience. Bertrand Russell cites a Sufi mystic that, 'past and future are what veil God from our sight'. The eclipse of the *it was* and the *it will be* leaves the *it is* uncorrupted. This collapse of time into duration, of the temporal into the eternal, and conversely, the 'descent of the eternal into time', transports the divine into the temporal, as Plotinus wrote, 'eternity as God manifesting his own nature',[17] an eternity that is not infinite continuance but absorption in the present. In many religions, a release from the convention of time entails a withdrawal from the world of temporal succession to a simultaneity of all past and future time, and from a duration that spans all of time it is a short step to a consciousness, the mind of God, that spans the arising and the perishing of the universe.

In Buddhism, release is also a liberation from the karmic cycle of existence and the cumulative efficacy of action. The descent of the mental state from present objects to past images (retrocognition), the importation of memory into the present, the unpeeling of personal memories in a direction from the recent to the archaic, memories of memories, may account for the intuition of karmic process. The elasticity of the past boundary of the present allows the present to expand not by a shift of its forward edge, which is fixed in the actual world, but by a recession of this boundary to a deeper, thus more distant past, with the result an anchoring of the present by archaic phases, finally by the inheritance of an animal memory from which those phases were derived. The access to the primordial stretches the present to embrace the ancestral, giving an awareness within the mind of an accretion or heap of prior forms deposited in serial layers, thus a concept of selves encrusted over lifetimes, recaptured in a cyclical return of appearances.

Time has a central role in the structuration of the categories of experience. Conceptuality runs through all things great and small. Duration is linked to the categories of phase-transition that constitute the smallest particles, no less than the comparison of phases that forms the limits of a present moment. Duration and conceptuality are thematic in mind and nature, ingredient in non-cognitive entities and explicit in consciousness. Duration is a thread of connectedness that can be traced from its culmination in human consciousness, to the origin of consciousness in the becoming of atomic entities.

Charles Hartshorne believed that the relation of mind to brain was analogous to the relation of God's mind to nature. Is duration the axis on which this analogy is drawn? The implication is that object persistence in the illusory consciousness of the mind of an individual samples the spatial continuum of a God's-eye view of eternity. We make contact with this perspective by realizing that all entities share the same process of momentary origination. Consciousness and the world are re-created in every mental state as change deposits epochs of time out of sheer simultaneity. That consciousness is a continuation of a trend that begins in material nature imbues life with sanctity. This is the meaning of objects that share in universal process. We speak of value in relation to conceptual feeling as the embodiment of sanctity in human acts and objects. The becoming of complex entities in the specification of acts and objects is the basis of their valuation. The insight of Whitehead's argument, and an implication of microgenetic theory, is an intuition of value in nature sensed through a sustained meditation on the connectedness of conscious duration and material fact.

Part III
Agency and Value

Chapter 7
Neuropsychology and the Self-concept

William James wrote that, '*a man's Self is the sum total of all that he CAN call his,* not only his body and his psychic powers, but his clothes and his house, his wife and children, his ancestors and friends, his reputation and works, his lands and horses, and yacht and bank-account.' James divided the constituents of the self into a material, a social and a spiritual self, and a pure ego.[1] The material and social components concerned the values, cultural artifacts, concepts, feelings, and physical substrates of the self. The spiritual self referred to inner subjectivity and the pure ego to a principle of personal unity.

This may be a promiscuous expansion of Kant's distinction of the empirical self and the self *an sich*, but it is a reasonable inventory of what a theory of the self is obligated to explain. What else could the self be but images, feelings, and experiential memories bound up in a momentary identity? The self-identical character of the personality over time and the sense of integrity for diverse experiences and feelings reflect the temporal and spatial unity of the self. The basis of this unity has eluded many fine thinkers.[2]

In the Buddhist or Humean philosophy of 'loose perceptions', the unity of the self is illusory. But if the unity of the self is an illusion, it is unlike other illusions that are perceptually discoverable. If the self is a bundle of images or a sequence of instantaneous flashings, how does the illusion of a substantial, continuous self develop? Identity at an instant or over time cannot arise from mere succession or association. Buddhism is mute on this point. Hume had little more to offer. He wrote that the problem of the connection among distinct existences, though not insuperable, was a difficulty 'too hard for my understanding', suggesting that one day some hypothesis will reconcile the difficulties. The problem of unity, or coherence, is the goal of any inquiry on the nature of the self, but first, what can we say about its relation to images, percepts, and memories from a neuropsychological standpoint?

The pathology of the self

Objects

Alterations of the self are prominent in disorders of perception, where the erosion tends to be in a direction leading inward from the object, that is, first the object breaks down, then the image, then the self. In cortical blindness with destruction of the visual cortex and a loss of visual objects, transient nightmares and hallucinations occur in the early stages. One could say that the surface of the percept peels away to expose the underlying image. After the initial confusion and hallucination subside, the self is agitated, anxious, and gives imaginary accounts of events, confabulations, yet some integrity of the personality is retained. A person with cortical blindness is generally unaware, or denies, that he is blind. The confusion is not the cause of the unawareness but a sign that the self is undergoing dissolution. The confabulation is pronounced for tasks that challenge visualization and may reach the point of a delusional psychosis. The condition is aggravated by depression when some light and dark perception returns, when the individual, no less blind than before, becomes aware of his blindness.

A similar pattern occurs with cortical deafness with bilateral lesions of the superior temporal or Heschl's gyrus, that is, confusion, agitation, auditory hallucination, and incoherence in the context of an acute psychotic break. A sudden disruption of sound or word perception with relative preservation of language presents as a psychosis. In both functional and organic psychoses, the boundary of self and world becomes fluid with a spilling over or merging of self into objects. Patients may be uncertain whether they are awake or dreaming. This may gradually resolve, but the self never regains its former character.

Comparable symptoms occur with disruption of somaesthetic perception in cases with a sensory paralysis, especially of the left side, and a lack of awareness for the existence of that side of the body. This condition is termed hemiasomatagnosia. The patients also show confabulation, denial, and hallucinatory substitution. Confusion is common, perhaps invariable. Although the consequences for the self of a lateralized defect in kinesthesis resemble those of visual and auditory deficits, proprioception may be more fundamental than exteroception, for it defines the limits and changing topological shape of the body as it moves in space. From an evolutionary and developmental perspective, the bodily sense plays a central role in subjective feeling.

In cortical deafness, paranoia is common. Whether it is intermittent or chronic, paranoia is not to be conceived as a delusion that the self exhibits, i.e., it is not a property or attribute of the self which, minus that attribute, remains unchanged, but is the sign of a delusional self, i.e., the self as a whole, for in that paranoid moment the self has been altered.

My interpretation of paranoia is that it represents the intrusion into wakefulness of a fragment of dream cognition in which the passivity of the self, which is typical of the dream state, assumes a persecutory mode as a victim to the events of its own imagination. The self is no longer the agent of thoughts and acts but a recipient of images and percepts. A verbal image shifts from an interior act to an exteriorized object. An example of this is when inner speech transforms from the verbal imagery of an interior monologue produced by the self to an auditory hallucination that is experienced as an assault.

The self can be partly sustained with the loss of one modality – vision, audition, kinesthesis – because a real world can be realized through the other perceptual channels. The self is less resistant to auditory than visual disruption, perhaps because of the involvement of language and the onset of delusional thinking. Delusion involves a derailment of thinking at a depth comparable to that of hallucination; it is a conceptual substitution in the linguistic mode.

When cortical deafness is combined with cortical blindness, and the auditory and visual worlds are lost, the self dissolves to a state of persistent confusion and disorientation. Cases of hemiplegic denial with cortical blindness show marked confusion. In a personal case the denial was selective for the hemiplegia and did not affect a prior finger amputation in the paralyzed limb. A subject with cortical blindness who had a prior blindness in one eye, denied blindness in the 'good' eye but admitted blindness in the previously blind one. This confirms that denial is modality-specific and specific to the acute deficit. When more than one modality is involved, the world partly disappears, replaced by an imagery that is sustained by the intact systems. In the destructuration, one could say that subjectivity reclaims the ground of dream and magical thinking antecedent to the division of mind into self and world.

In sum, as well as can be ascertained from the presence of confusion and delirium, the clinical material demonstrates that visual, auditory or kinesthetic perception cannot be sacrificed in the mature individual and the self remain unaffected.

The self also decomposes in such conditions as snow blindness, when the visual field becomes uniform, object contours dissolve, and the horizon disappears. But in snow blindness, or for that matter, when we close our eyes, or in acquired peripheral blindness, the visual world is still tacitly there. This is also true for peripheral deficits in hearing and proprioception. In cortical blindness, cortical deafness or hemiasomatagnosia, a portion of the world no longer exists, not conceptually, not in consciousness, not in memory; rather there is an emptiness in the domain of that modality.[3]

The signature of the altered self is confusion, with impaired attention, restlessness, and incoherent speech, but what exactly is confusion? Its pathological substrate is unknown (see the review in Seltzer and Mesulam

(1988)). The central defect is the impairment of attention, but that is another way of describing a disruption of focal or analytic perception. Focal attention is focal perception, not perception focused by attention. Although a lability of attention can occur without confusion, as in children with attentional disorders, or in frontal lobe cases, confusion does not occur without a disorder of attention. In the former case, there is distraction from one object to another within a relatively stable perceptual field; in the latter, attention shifts with the replacement of objects, perhaps through a lack of selective interest or affect, i.e., affective tonality is not fully differentiated, or is distributed over the ambient field.

An alteration of the self entails an alteration of subjective time. Objects are the bearers of temporal relations that articulate the field of consciousness. Objects are like time-lapse photographs, except it is not the slow change that is filtered out but the rapid change that is concealed. With a loss of objects, there is a loss of the continuity of events in subjective time. When this is lost, the present contracts and the self and world are in danger. The causal succession of objects that was formerly concealed in the glue of the present is unconcealed, with the result that there is an accelerated replacement of objects in durations that are foreshortened, thus distraction, rapid shifting of perceptual content and an instability of the focus of attention.

Imagine a present no greater than the blink of an eye; each time the eye opens, a new world appears. The fragmentation of the world has its correlate in the fragmentation of the self. A person who cannot see a chair as continuous is not a continuous person. A similar phenomenon occurs in severe amnesia when forgetting encroaches on immediate memory, such as that displayed in the digit span or in retaining a telephone number. This occurs in severe postencephalitic amnesia or Alzheimer's disease. The psychologist David Wechsler, when in his 80s, eloquent and conversational, described to me his sense of isolation and fragmentation of the ego as a 'loss of the thread or continuity of the self', when memory loss began to invade the immediate past. When the duration of the present contracts through rapid forgetting of the immediate past, consciousness does not extend over a duration sufficient to enclose a succession of events, and the self lives closer to the knife-edge of physical passage.

Such observations would be unintelligible if the self were to be conceived as a passive recipient of the external world, for then the world could fly apart and the self remain intact. If objects are assembled in the cortex and associated to the self, the self should resist perturbation when the visual cortex or its associative links to the self are destroyed. A change in the self might then be construed as an emotional response that would vary according to the personality. However, the picture I have described is systematic across and within subjects. Such observations support arguments by Kant, Hegel, and others that the subject-object relation is fundamental. Self and world are not separate entities that interact, but relations in a continuum of transformation. The world is a product of the same process that lays down the self.

Hallucination

If we turn from changes in the self with object breakdown to the changes that occur with hallucination, we see a similar pattern. A visual image, say a face, may be apprehended as hallucinatory until it is combined with an auditory hallucination, say when the face begins to speak, at which point the image is perceived as real. A hallucination in two modalities tends to be apprehended as a real perception. It is important to note that the image does not become more objectlike when it becomes more real. The hallucination is not accepted as a real object because it looks like an object, it is accepted as real because it is the only object the person has. There is no other object or image that is truer to reality to which the person can appeal. When a hallucination coexists with a perception, for example, when a hallucinated head is visualized on a pair of real shoulders, or when a hallucination coexists with a real object in a different modality, its hallucinatory nature is usually recognized.

When hallucination replaces perception, there is no real world left in that modality. When hallucination invades more than one perceptual system, it becomes the only world the perceiver has. In dream, or in wakefulness, there is a conspiracy of the senses so the individual cannot disconfirm the imaginary nature of his images. The world is accepted as real when its reality is not disproven by one of its modes of representation. Dr. Johnson's fatuous refutation of Berkeley by kicking a stone shows just how inadequate common sense is to these issues.

Over time, with chronic hallucination, there is the formation of delusional beliefs. The schizophrenic who hears voices that he believes to be real does not need a visual hallucination of a speaker to reinforce the belief in their reality. Auditory hallucinations are more likely to be accompanied by psychotic behavior than visual hallucinations (as confusion is more severe with loss of auditory objects), partly because a real object in that modality usually does not occur simultaneous with the hallucination, the reality of a hallucination in one perceptual system requires a delusion to support it, and delusion is more closely related to the auditory modality.

In autoscopy and out-of-body experiences, the image is an approximation of the public self similar to that of ordinary hallucination except the subject is the content of the image. The image is like a mirror construct of the outer appearance of the face or body often, as in hypnagogic hallucination, with more vivid coloration and emotional expression than in life. An interesting feature of out-of-the-body experience is the cycloramic quality of the self-perception. Autoscopy is associated with frontal and temporal lobe pathology, but it has been described in a variety of altered states. The poem by Heinrich Heine, *Der Doppelgänger*, is an example.

Da steht auch ein Mensch und starrt in die Höhe
Und ringt die Hände vor Schmerzensgewalt,
Mir graust es, wenn ich sein Antlitz sehe
Der Mond zeigt mir meine eigne Gestalt.[4]

The hallucination of the self is the other side of a failure to recognize
the self in a mirror, the so-called mirror sign of severe dementia. Such
patients may fail to recognize their reflection, though other faces are
identified. There is also a loss of mirror space; i.e., if one introduces an
object behind their head, they reach into the mirror. Of interest, they may
not show the ability, which has been demonstrated in the chimpanzee, to
apprehend a marking on their own face that has been surreptitiously
applied and noticed in the mirror image.

If we compare the self in object breakdown to the self in hallucination,
we find that in the former, an intact modality supports the self, whereas in
hallucination an intact modality supports the distinction of what is real
and what is imaginary. In either case, whether the destructuration of the
world that erodes the self or an erosion of the reality of the world that
destructures the self, the self and the world are realized largely through
visual, kinesthetic, and auditory perception and linguistic derivations, but
neither self nor world can be sustained when more than one modality is
affected. The reality of the world is linked to the integrity of the self. A real
self does not feel that an imaginary world is real nor that a real world is
imaginary.

In psychosis, the disintegration of the self may begin in an object or a
personal memory, for example, when an illusory whisper becomes a hallucination of voices or when a memory image is 'seen' as a hallucination.
The incomplete object is accompanied by derealization or loss of reality.
The incompleteness entails that the object retains a portion of the subjectivity out of which it developed. The preliminary in cognition has a greater
share in the final object. The object is more like a thought, which is to say,
its cognitive origins are felt because the object is not fully objectified. The
feeling that is apportioned to the object, that normally deposits in the
object as value and the sense of reality, is now attenuated in its outward
trajectory. The retraction of feeling deprives the object of realness. It
becomes lifeless, mechanical.

From such observations we conclude that the belief in a real world
does not depend on the realness of the perception but on the coherence
of perceptions within and across the different modalities. The integrity of
the self and the feeling of the reality of an experience do not follow a
judgment of a correspondence between mind and world but an agreement across perceptual modalities within a momentary act of cognition.

Agency

We usually think of volition in relation to action, but the action we feel is a
corollary perception of a movement for which we are otherwise unaware.

William James wrote, 'in perfectly simple voluntary acts there is nothing else in the mind but the kinaesthetic idea'. Action contributes the active quality without which there would be passive movement. This is the 'feeling of innervation' or self-initiation that was at the heart of the debate between James and Wundt. There is also a relation to will and drive and the outward distribution of affect. Action inheres in every percept, but agency as an intentional act seems largely a relation of the self to the *perceptual* aspect of a cognition, whether a motion or an image. For this reason, agency can be examined in relation to imagery and perception as well as to motility.

The self is active in thought or imagination images. One can visualize an elephant dancing on the head of a pin. There is a volitional quality and a feeling of control. This is less pronounced for the memory image, but still there is a feeling of volition in the effort to search a 'memory store'. We feel that we scan an object that we term memory and that this object, memory, is distinct from the act of searching, i.e., that memory is the target of a search, not the instigator of what the search discovers, when in fact the feeling of agency that accompanies the search is produced by the conceptual phase in 'retrieval' that is striving toward expression. There is a quality of agency even in a dissolving eidetic image, at the point of fading, where an attempt at revival discloses a transition to memory. In contrast, the perceptual object is felt as independent of our actions. There is a progression from images that come unsolicited, as in dream and hallucination, where the self is passive to the image content, to images where the self feels an agent, such as thought or memory images, to images that detach as external objects, or perceptions. If we conceive the different forms of imagery as moments or segments in the process of object perception, and if the self is generated in this process, as I believe the neuropsychological material confirms, we can interpret the feeling of passivity, agency, or detachment as specific to a phase in both self- and object-realization.

Agency and language

Generally, agency is expressed in terms of its intentional, deliberate quality, in the ability to direct or withhold an action, and in the awareness of choice. An aphasic with a residual word or phrase, or a patient with echolalic speech, produces words in a rapid-fire manner without choosing which words, if any, to produce, and without evidence of awareness for the words that are produced. There is a correlation between the volitionality of the act and the consciousness of its content. This changes each moment, so that deliberate utterances with awareness may alternate with automatic ones without awareness. If agency and awareness fluctuate in concert, the self is part of this fluctuation. The self of the moment is driven by the momentary state. One concludes that the momentary state of self, agency, and consciousness and the degree to which the performance is realized are interdependent phenomena.

In the acquired disorder of deep dyslexia there is less awareness for semantic than for visual or phonological errors. If a person reads the word (or names the picture) horse as 'zebra', he is less likely to judge this performance incorrect than if he reads the same word (or names the picture) as 'house' or 'hort'. The semantic is early in process, generally less conscious than the phonological. Cases with semantic errors show less awareness of deficit than phonological cases who have error-awareness and self-correction. On semantic priming tasks, total aphasics with phonological deficits are able to extract word meaning without awareness of the words. They show inference without evidence of word understanding. Thus, given a sentence such as 'John forgot to close the door', and shown a picture of a closed and open door, they point correctly in spite of their apparent inability to read a single word.[5] In other words, consciousness is material-specific with regard to the semantic and phonological phase of the cognitive process. This specificity corresponds with levels in the realization of language from an antecedent phase of semantic processing to a consequent phase of phonology. The self in a semantic state has a different quality of consciousness from the self in a phonological state. The pathological material helps to dispose of the notion that the self is a substance and mental states are its properties. The properties of the self are states of its existence, not changing features of a self that is unchanging.

Unity

With all this by way of introduction, we can now ask what is the source of the unity of the self. The self cannot be a mere bundle of impressions, for the feeling of unity across successive states differs from the feeling of change across successive bundles. The self is the same from one moment to the next, but ideas and impressions are constantly changing.

Spatial coherence would seem to necessitate synchronicity; the activation of *elements in phase* is essential for them to be included in the same state. But simultaneous activation does not explain unity; the world is synchronous in its passage. Self-identity cannot derive from linear coherence which entails an open-ended continuation with no possibility of comparison (e.g., past/present; figure/ground). Contrastive features are important. Contrast involves relations which implicate temporal process. In the relation of figure to ground, there is a microtemporal transition in the kinetics of synchronicity that carves local unities out of the simultaneity of the whole. The unities are replicated – the self is more or less replicated each moment – but the replicates must be discontinuous if they are to be compared. I have argued for a cyclical recurrence of brain states as atomic unities, with slow change of configurations at the depth and variation at the surface, i.e., a core self that is derived or 'speciates' into a changing array of mental contents.

For William James, personal identity was a judgment of sameness across instants of thought. Thus, given a self of yesterday and a self of today, James wrote that a pulse of cognitive consciousness not only thinks them both, but thinks that they are identical in spite of the fact that each pulse of consciousness dies away and is replaced by another. The difficulty with this interpretation concerns the judgment of sameness, because the replacement is a revision in which the prior state perishes in the act of being replaced. There is no static content corresponding to the self of the preceding moment that can provide an object for judgment. If the present self were radically different from the prior self, would the present self know or sense this difference? Is there awareness of difference at all, if the present self is the only self that exists?

James argued that the appropriation of the past thought by the present thought occurs by the arising of the latter over the residue of the former, with each pulse of consciousness containing the prior pulse within it as an overlapping wave (see Figure 7.1).[6] A thought has access to a prior thought that is embedded in its structure. James did not explain how the states were compared, nor the relation of the past to the present state in the duration of the present, but Köhler, in 1923, suggested that the duration of the present develops as the comparison of a fresh impression with the 'fading trace' of the prior state (Köhler, 1923). I would say the fading trace establishes the posterior boundary of the present, while its anterior boundary is formed by the actuality that is deposited. A past thought revived in consciousness is a present image attributed to the past by its incomplete recall. The incompleteness gives the image its pastness. Were the image fully revived, it would be experienced as a present object, such as a hallucination, or a perception, not a memory.

The self arises at that phase of the mental state corresponding to the posterior boundary of the present, or the floor of the immediate past, into which each fading act of cognition returns. The self is 'positioned' in a transitional segment between the unknowability of the physiological unconscious and the facticity of the known world. Because of its antecedent 'location' and its relation to the pastness of emerging or

 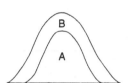 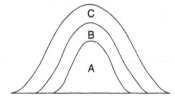

Figure 7.1. For details, see note 6.

decaying content, even in the immediacy of a concrete actualization, images and objects are perceived as a continuous advance into novelty.

Hume examined the self and found only images and impressions. James found physical sensations. The potential of the self to generate ideas and impressions does not yield its subimaginal content even to the most powerful of introspectionists. That is because the self is not a collection of elements that compile to a sum but a configural antecedent of the elements into which it is derived. Thought thinks up its images in such a way that they are felt to be products of the thinker, and thought thinks up its objects in such a way they are felt as independent. In truth, however, self, image, and world are a single object in transition.

The duration of the present creates a theater for experience, but does not fully explain the unity of the self. We need also to explore the nature of categories and the analogy of a duration and its instants to a category and its members. A duration is a 'container' of temporal parts; a category is a 'container' of spatial parts. A duration is a container of arbitrary length that enfolds a number of instants; a category is a container of an arbitrary set of members. A category is like a duration in that both are enclosures with fuzzy boundaries for virtual parts that are, themselves, potential containers. An instant is like a member of a category in that it can be a category for another member. Duration is the primordial manifestation of categorization.

If duration is a kind of temporal category, and an abstract class is a kind of spatial category, could this account for the unity of the self 'at an instant' and the self-similar nature of the personality over time, i.e., temporal and spatial coherence? The changing states of the self, the Humean bundles of instantaneous feelings and impressions, would be subsumed in the categorical self that is revived each moment, while the surface contents of the state disappear and are replaced by other contents. Unity and identity across successive occasions owe to the reappearance of a categorical self that transcends its momentary realizations.

Process and category

In prior writings, I urged a view of consciousness as the relation of the depth of the unfolding mental state to the surface actualities (Brown, 1993). The antecedence of the self in the phase-transition guarantees a feeling of causal priority in relation to action and a feeling of autonomy and opposition in relation to perceptual objects. The self was conceived as a preliminary segment of unspecified potential from which acts and objects emanate. The description emphasized the insubstantiality, or relationality, of consciousness and its dependence on the relation between successive planes of space formation.

We are now in a position to argue that the relation between depth (self) and surface (image, object) in the mental state, or before and after in microgenetic process, with the self at the posterior boundary of the present, is transposed to a relation between an observer and an entity at either limb of the duration of a conscious moment. The relation between the phases, i.e., the before and after, the relation of self to world, extrapolates to a now in which phases solidify as fictitious components. The relation of succession in a nontemporal (vertical) becoming transforms to one of precedence in the serialization of the (horizontal) present by events. The nontemporal actualization precipitates cotemporal entities. The concurrence of these entities in the specious present is consciousness, i.e., the conscious state. The self forms the past boundary, the surface images and objects form the actual boundary, of this duration. The core revives the self and the past, which are specified by sensory experience to perception. Conversely, the decay of a perception leads depthward to the past of memory, to the self and its personal past and beneath that to what remains as character after experience has been forgotten.

Consciousness

The self is ingredient in consciousness. But how is consciousness to be conceived given the multiplicity of assumptions regarding its nature: a device for self-monitoring or regulation, a computational algorithm, goal-direction and intentionality, or the stuff from which the universe is created? Perhaps we should first ask what the conditions are that need to be satisfied for a mental state to be conscious.

A self is necessary; there is no consciousness without an observer. The absence of a distinction of self and object eliminates the subjective by a law of parsimony. A perception is necessary, for there is no consciousness without an object and an object space. A state of consciousness cannot be long maintained with a loss of perceptual objects. An immediate past is necessary for a comparison with the present, as is a duration spanning the comparison which the self and its objects inhabit. Without a duration, an act of cognition would be a slice of infinitesimal thinness on the forward edge of change. Self, world, and present, mutually penetrative and interdependent, are the essential ingredients or relations of consciousness.

If the duration of the present is specious or illusory, as William James maintained it must be, and if consciousness develops in the context of a duration, as any reasoned analysis must conclude, consciousness (the conscious state) is *speciously* functional. If there is no entity consciousness, there is no function or localization. If consciousness is the state of its constituents, the nature of consciousness will be determined by the state of each constituent. This includes the duration in which each constituent develops.

Duration and illusion

If the self and its conscious duration are illusory, at what point does the illusory nature of categories become the conceptual nature of reality? A category is an abstract class of entities or, in process theory, the abstraction over phases in a single entity, but in any case not a thing one can point to and say, this here is the category. In the becoming of a mental state, the initial phases that deposit the self are not extinguished by the time the final phases are reached but are present with the final phases to give a duration in which the self and the object constitute a momentary whole. The self at the core of the mental state remains alive with the actual surface to form a whole entity of mind and object. This is possible because becoming is nontemporal. A whole entity, a fully unfolded mental state, has to be achieved before there are temporal facts.

The world is conceptual to its limit. The ultimate bits of nature are categorical in the epochal nature of their constituent phases. The conceptuality is intrinsic to the object. A category is no less real than an object. Categories are the generic in process through which diverse entities are unified. This is also true for the self. A reality that is categorical is perceived as unreal only by those for whom the real is irreducible.

World and mind are phenomenal, but their phenomenal nature is not the basis on which their material ground is to be rejected. Some phenomena are more illusory than others. Some illusions we notice, such as a Müller-Lyer illusion; others are discoverable, such as object constancies. On other illusions, people disagree, such as the illusion of time. Still others, such as the hallucinatory nature of the self and perception, defy reason and, when the illusion is invaded by conviction, invite a schizophrenic collapse.

An illusion is an image that does not correspond with reality. A hallucination is an illusion that deviates so far from the reality of perception that it seems to be another reality. Illusions are perceptions that digress; hallucinations are digressions to the point of novel percepts. The difference between hallucination and illusion is the degree of realization. What makes a perception real and a dream hallucinatory is the idea of a separate reality to which only the former corresponds. This way of thinking reflects a view that goes back to Descartes of a formal existence for mental images and an objective existence for physical ones. In illusion or hallucination, there would be a discordance between the two worlds.

The extent to which the world objectifies determines what kind of world is produced. Generally, a uniform image of an actual world is not accompanied by a memory that is more real than the one that actualizes. The apprehension of a discordance may come later when a perception appears 'more real' than the illusion. Those who sense that the world of perception is illusory may have a vision of another world that is more real. However, the *lived* feeling of discordance only occurs when an incomplete phase in mental process, such as a memory image, is embedded within a conventional percept.

The universe at a moment includes the process in every brain. The self that is generated by the brain, or the brain state of a subjective experience, is both a perspective on a world as well as an object in its composition. The mind is a process of objectification that deposits different momentary selves and worlds. The world deposits the mind as one of its momentary objects. A brain actualizes as any object in nature, becoming what it is for that moment, i.e., a behavior, a mental state, to be replaced by another actuality. We do not ask whether the actuality of an objective entity, say a tree, corresponds with the rest of nature since the tree is a natural object like any other. Noncognitive entities are what they are each moment of their existence. We do not compare one tree with another and say that one is more real. A mature tree has no greater claim on reality than a juvenile shoot. Nor would we compare in this way the stages in the life of an individual. We can think of the mind in the same way. An illusion is the result of the failure of the brain state to reach a certain maturity of momentary actuality. A mental object is illusory or veridical, i.e., incompletely or fully actualized, in comparison with a world object from another perspective. The mind (brain) is an object in the world at that moment. There is no comparison involved.

The brain state realizes a momentary cognition as an object in nature. The perception of nature that fills that cognition displays the brain state, not the field the perception happens to depict. The field is a model. The reality that is modeled, and the correspondence of model to reality, is inferred from the pragmatics of survival. In a strictly pragmatic approach, however, the degree of approximation is determined by the accuracy. Because the only reality a perception could depict is one in which the organism could survive, the argument from utility or adaptation is circular.

Correspondence and adaptation

The interdependence of components in consciousness, and the dependence of a perception of the real and the integrity of the self on a concordance across modalities, raise the question of coherence generally both within and across conscious states. Whitehead (1948) attributed the correspondence of the mental with the physical to the establishment of a dominant space-time through a continuity across successive states of waking awareness. Each mental state is a contingent phenomenon that is dependent on the momentary state of awareness. The continuity of states across shorter or longer intervals reflects an awareness of the self-similarity of the space-time continua of the states. The coherence across the states is evidence of their conformity to an external space-time that is presumed uniform – that is what accounts for their similarity – and this conformance, the 'internal fit of speculative concepts and empirical world' (Bradley, 1994), is the basis for our awareness of reality. The notion of conformance is not innocuous. It seems to involve the very bifurcation of mind and world that Whitehead deplored.

Whitehead further argued, in line with ordinary experience, that the continuity from one state to the next and the fit of a single state with the process of nature is lacking in dream. An awareness that is discontinuous and does not fit with the dominant space-time is termed 'imaginary'. But the mental state always conforms to the brain state. We do not believe that mental states have their origins elsewhere than in brain states. We have no way of knowing whether the brain (or mental) state conforms with nature external to the brain nor even what it means to say that the process of brain activity does or does not correspond with the process of external nature. The brain state is always in conformance, for it is part of nature. With a lack of conformance, the mental state is in compliance with the brain state, but one portion of nature, the brain state, does not correspond with the nature modeled in perception. To apprehend a fit of the mental state with external space-time is to make a comparison, yet the space-time of nature is imperceptible.

If a mismatch between one's perceptions and the process of nature is imperceptible, a mismatch between the perceptual components of a single mental state in the normal individual is often barely discernible. We take for granted that visual, auditory, and kinesthetic perceptions will be 'in synch' and realize the same world. The reality of the world depends on this correspondence. The reality of an experience is not conditioned by a fit or misfit between the continua (waking perception or dream with external nature) but by a correspondence across the constituent modalities within the mental state, i.e., by an incongruous actualization of terminal entities in the brain state.

Thus, an individual may know that one perceptual modality does not conform with another, but he cannot determine whether the mental state conforms to external nature. One can give reasons, when awake, for believing that dreams are hallucinatory, but the dream experience is usually felt as real *in the dream*. There are lucid dreams in which the reality of the dream image is questioned, but there are also conscious states in which the reality of waking percepts is in doubt. In dream, we may question the reality, when awake, we may wonder if we are dreaming. For some, the uncertainty as to what is real is more terrifying than the prospect of an unreal world that it throws in question. A mild bout of vertigo throws me into an epistemic crisis. Reality is in danger when we feel the conformance or adaptation of the mind to the world, for then we sense that the world is an image. We sense this because we feel the lack of coherence among our own percepts. The moment we experience a match or a mismatch between entities in the mental state, we become aware that objects are mental pictures and a wedge is introduced between the observer and his world.

The feeling of (non)conformance across concurrent perceptions reflects a disparity in the worlds that actualize. An incomplete percept, say a disrupted visual perception, actualizes a preliminary object. This object is out of step with other perceptions and has the quality of an image in the immediate past of the observer. The incompleteness of the actualization is comparable to the incompleteness in the revival of a prior percept. An object that is not fully realized may resemble one in decay. An image that does not achieve perceptual clarity and presentness resembles one that decays to the vagueness and pastness of imagery. An image conveys a world that is neither exteriorized nor in the present, rather like a dream one tries to hold onto. A comparison with ideal nature is the basis for a theory of correspondence, but the comparison is indirect. The real nature of the brain state actualizes in every perception.

In dream as in wakefulness, we are unable to compare an image to an external space-time, but in dream we cannot evoke comparative judgments of prior episodes, though a progression and recurrence of incidents may occur over serial dreams. There are recurrent dreams but recurrence in waking is *déjà vu*, not coherence. Why do we recognize a continuity of conscious states within consciousness and not of dreams within a dream?[7]

I think the lack of continuity in dream is due not to a lack of conformance but to the absence of a past, a present, and a future. The now of the dream has a limited temporal thickness. In waking consciousness, the past is implicit in anchoring and determining the present content and explicit when it is revived in the imagination, for example, as a memory image. In the dream, the now is not sandwiched between a past and future. There is a global happening in which the image is the total content. It is judged as past only afterward, on waking, when the dream is buried in an actual present.

In this respect, the space-time of dream more closely approximates our concept of physical space-time than does the space-time of consciousness. In the dream, as in the material world, there is a before and an after, but no past and future organized about a present. A dated past requires a now in which a past occasion can be compared with an occurrent state. It is even questionable whether the sequence of dream events is apprehended as a sequence in the dream or is achieved on waking. The possibility that seriality and temporal order are realized on waking out of a simultaneous dream content is discussed in Brown (1993). Dream cognition is unique in other ways that reveal, by its absence, the contribution of a durational now. The dream is a fluid avalanche of moving events unaccompanied by a critical or discriminative awareness, nor a feeling of agency. The self of dream is drawn by the events to which it tends to be a passive witness, often a victim of the acts and images that occur, unable to pause, choose, judge, or veto a course of action.

Consider a series of conscious states: do we invoke a past experience, compare it with a present experience and label the two as matched or mismatched to each other and, individually, to physical space-time? What is the standard for incomplete or inaccurate recall? For a recollection to be judged as inaccurate, there must be a more accurate version in memory. A past event in awareness is an image in the present that does not reach the phase of a concrete object. The image is less vivid than a perception. The lack of concreteness and clarity are part of its 'locus' in intrapersonal space and past time. When a memory becomes vivid, it may actualize like a perception, for example, as in hallucination. The feeling of the reality of an entity depends on the completeness of actualization in waking cognition or on the incompleteness of all perceptions. A comparison of a past and present state over successive moments requires a duration, if for no other reason than to separate the events. A mental state is momentary, so a comparison across two states is a comparison in the momentary state of the limits of an illusory interval.

To summarize, the ability to judge a coherence over successive space-time moments of consciousness may be a criterion of reality, but it is not the grounds for a judgment of what is real. An act of cognition cannot be compared to physical space-time, so the judgment of reality must depend on a comparison across successive conscious states. The comparison is between the degraded image of the immediate past and the knife edge of an actual present. We do not have concrete images of the past to compare with those of the present. The judgment of reality depends on the present state. The *feeling* of a comparison is possible in waking consciousness but not in dream. This feeling is a sign that an adequate (for survival) version of the real has been achieved. The duration of the present is the medium in which this comparison occurs. In cases of brain pathology, a degraded perception can be the basis for a judgment of the unreality of the world through the lack of perceptual coherence.

Having created an external object, we are compelled to employ it as a criterion. What would be the character of correspondence if all our objects were personal? The judgment of an accurate model, or a fit of our concepts with an external space-time, must be reinforced by the coherence of components within the state. If one runs from an imaginary tiger and ignores an oncoming train, there is a lack of correspondence, but to the observer, not to the subject who perceives a tiger that seems real. The perception is anomalous but does not feel imaginary. Anomalies do not survive, whether organisms, species, cognitions, or theories. The organism fails to respond to what is exigent in the environment, but the organism is part of the environment and perishes for a lack of adaptation, not for a lack of correspondence, which is inferred from an unsuccessful coherence.

When we leave consciousness and descend into the foundations of the mental state, we recover the objects of nature out of which consciousness is elaborated. The category of consciousness differs from that of objects in that it includes all object categories as members. Consciousness is a category in which self and object have a share. Self and object are members of the category consciousness. The real is the coherence of these members at successive phases in the striving toward definiteness. The self is part of the coherence and must be as real to objects as they are to the self.

Finally, to sum up what this signifies for the biological basis of the self, the neuropsychological material demonstrates that the self is deposited in the process of object realization, that it distributes into images and objects, and that a truncation of this process results in an erosion of the self that is similar across the different perceptual modalities. The self is categorical and relational, achieving autonomy in the context of a complete derivation. The autonomy depends on the completeness. The preliminary locus of the self in the mental state entails a holistic or multimodal phase of potential prior to perceptual individuation. This, together with a relation to feeling, to the personal history and the immediate past, point to a limbic transition in the outward development of the mental state.

Chapter 8
Subjectivity and Scientific Thought

Every day we read of a new scientific discovery, a photograph of one of Jupiter's moons, a computer that composes Bach, a gene for a rare disease. The sense of forward progress fills us with admiration. Why is it, then, at a time and in a culture where scientific thought is so dominant, interest in quackery has never been stronger? More people in this country believe in extraterrestrials than DNA. Indeed, many educated people – even the most accomplished physiologists and philosophers – have beliefs that are completely at odds with the rest of their knowledge base. Think of Fechner's spirit-writings, Eccles's psychons or Axel Munthe's touching description of William James sitting for hours at the bedside of a dead friend, pen in hand, paper blank, waiting to receive messages from the beyond.

The most common interpretation of this phenomenon – the compatibility in a single mind of what seem to be incompatible belief systems – is that a scientific picture of the world is impoverished if it excludes the subjective, and that, if the expression of subjectivity through science is blocked, another path will be found, in the magical, the occult or the religious. The prevalence of such beliefs taps the popular conviction that there is more to life and experience than science can explain. However, what is unexplained by science does not begin at the limits of what is provable, which is the popular justification for magical belief, but rather, within the very objects of its knowledge. The incompleteness is apparent when the simplest question is asked, for example, if we ask, *what is an object*? The response of science to this question is that an object is an existent that does not change over brief durations or that the change in the object is cancelled by inter-subjective verification. The depiction of an object as a solid interacting with other objects, whether rocks or particles, reinforces a concept of *nature-at-an-instant*, where the relations of objects are extrinsic surface contacts. In contrast, the inner dynamic of an object, its temporal existence or internal relatedness, the *process-within-nature*, or the activity behind the surface of an object appearance, is the 'locus' of its quality or 'essence', in the sense of an intrinsic non-substantial nature. The often simplistic beliefs and protestations of the proponents of pseudoscience tend, like science as well, to ignore this interior dynamic, which is the source of the subjectivity of an entity, while the vague intuition that this dynamic is neglected in scientific accounts of mind and nature is the tacit ground for public dissatisfaction with the scientific mode of thought.

The crisis in science

The crisis in science is part of a larger problem that can be traced to the rise of positivism in the late nineteenth century. Collingwood restated the positivist manifesto nicely in his essay on historiography: 'each fact was to be regarded as a thing capable of being ascertained by a separate act of cognition or process of research (and) . . . each fact was to be thought of not only as independent of all the rest but independent of the knower, so that all subjective elements . . . had to be eliminated'.[1] In both mental states and natural objects, an entity was visualized as a spatial solid with a causal effect or output in relation to other objects, for this was what could be verified, across measurements, across perspectives and at different times. In psychology, behaviorism moved inward, postulating little black boxes within larger ones, each having its own input and output. This line of thought has led to the concept of the mind as an abstract computational space of non-relational solids. Whitehead wrote that positivism takes the attitude of a 'complete contentment with an ultimate irrationality'.[2]

The ascendance of a relational physics has had a negligible impact on cognitive theory, where mechanistic accounts of physical interaction continue to be the starting point for theory construction.[3] This is true for innatist models, in which the mind is read off the genes, as well as for experientialist models, in which the mind is an acquisition of sense data in patterns of association guided by innate mechanisms. Genes and sense data are determinants of behaviors, one intrinsic, for behaviors that are released or unlocked, the other extrinsic, and the basis of learning. In both learned and innate behavior, mental content is the result of an assemblage of functional parts that can be studied in isolation like other physical objects. The context of a mental or neural element is conceived as a non-temporal aggregate that can be trimmed away at no expense to the element, since the temporal relations between parts, being extrinsic and spatial, can be sacrificed without loss.

The elimination of intrinsic relationality (internal process) and the trivialization of the subjective have had profound effects in all areas of life and thought. An example is the description by Arthur Danto of the impact on him of the exhibit of the *Brillo Boxes* at the Stable Gallery in 1964. In Warhol's work, the repetition and banality of topics are designed to eliminate meaning in the image. Danto wrote that this work illustrates the fact that 'two outwardly indiscernible things can belong to different, indeed to momentously different, philosophical categories'.[4] In the philosophy of art, the distinction of the genuine and the artificial, the original and the copy, has traditionally been decided on the subjectivity – the meaning, feeling, intentionality – experienced in or imported to the artwork, but the question raised by these paintings, what is a work of art, is more important in many ways than any potential answer, for the question, as Danto pointed out, liberated art from preconceptions of what a work of art should be.

In neuroscience, the distinction of the authentic and its replication is the legacy of the Turing test. If it walks like a duck and quacks like a duck, even if it thinks it's a dog it's still a duck. The *as if* of the simile becomes the *is* of identity. The resemblance of human cognition to some aspects of computation provided the predicates needed for the topic of one argument, the computer, to replace the topic of the other, the brain.[5] The dissimilarities between the authentic and the replica are trivialized and the shared properties become the vehicles for an identification of the objects they modify. The principal difference that theorists acknowledged, that of hardware, one made of silicon, the other carbon, was judged to be irrelevant. Other differences between the original and the model were ignored, and the task of the psychologist became that of fine tuning an approximation to mimic an ideal.

In the philosophy of mind, as in art, the distinction of the organic and the model is less important than the opportunities it affords for theory construction by the perplexities it generates. But, as in art, we are confronted by the problem that the complexity of the critique can no longer be distinguished from that of the material to which it is applied. The model began as a heuristic, an artifact, then became a challenge in its own right, and eventually replaced the original.

In art, the freedom from tradition and the relative lack of constraints on style resulted in an explosion of innovative form. In cognitive science, the question, what is a mind, was largely rhetorical for the answer was already at hand, namely that the replica was equivalent to the original. The question did not liberate psychology for a period of creative advance but channelled invention into the same (dysfunctional) family of ideas. The progress was in the model or replica, not in the system that was replicated. This explains the indifference to translation from computational to clinical descriptions. The effort went into a correspondence of the computational model with the neurophysiological hardware, not with competing accounts of natural behavior, and the more complex the model became, the more remote was the possibility of translation. In psychology and in art, the question, what is a mind, eroded prior assumptions on the nature and extent of mind, and destroyed the narrative base on which those assumptions rest.

The issue simmered while cognitive science forged ahead but then, as now, the debate turned on the distinction of a conscious subject from an automaton or zombie. In spite of a century of protest, most scientists still live on zombie island.[6] The zombie, in fact, is no longer a science fiction creature but the prototype of the modern man. There is even a condition in which people have the delusion they are dead, i.e. zombies. In fact, some philosophers may be victims of this disorder, for they pretend to be zombies, at least we think they are pretending. To use the title of Francis Crick's best seller, this is really an astonishing hypothesis! It seems we must decide if subjectivity matters before we can hope to understand it.

Magical belief and scientific fact

The impenetrability of science to subjectivist doctrine is a mere frustration compared to the dismissive attitude toward (the significance for cognitive theory of) religious and primitive thought. Magical theories in science such as cold fusion do not erode the authority of the scientific method to the degree that superstitious beliefs undermine the import of primitive thinking. I do not wish to imply that false scientific beliefs should weaken our confidence in the scientific method, nor do I want to promote magical thinking as an antidote to science. As Dean Inge wrote, 'pragmatism is defenceless against obscurantism'. It is the *pattern* of primitive thought, not the content, that deserves closer attention. Thought is an engine of knowledge and its content can be appropriated independent of its means of production. But, the laws of thought cannot be extracted from the entities they actualize, no more than the function of a television set can be explained by looking at the picture on the screen. A thought consists of the content *in addition to* the process through which the content develops, i.e. the immediate history of the thought is as much a part of the thought as the final 'output'. A description of this process, in which the primitive foreshadows the religious and scientific, can provide a basis in cognitive theory for understanding the foundational laws of mind.

The conventional view of superstition is that it is a system of undomesticated beliefs that, so it is claimed, tend to cluster at the fringes of the knowable. On this view, ignorance or *uncertainty* is the garden where the primitive grows like weeds. The removal of knowledge, as Kant said, leaves room for faith. However, ignorance is not an obligatory feature of magical thinking, which is universal even in the well informed and highly educated (see below). To say that magical thinking stems from a lack of knowledge is the pejorative grounds for dismissing it as nonsense. The trivialization of the irrational motivates another interpretation, that it represents a *reaction* to the narrowness, elitism and dehumanization of science, what Heidegger called the 'emasculation of the spirit'.[7] A reaction, however, points to a mystery it does little to clarify and, again, is not a justification. A want of knowledge and a reaction to its limits play a role, no doubt, in the evocation of the magical, rather more as conditions of its appearance than explanations of its cognitive role. The irrational is important because the ubiquity of magical belief confirms what we already know from psychoanalysis and poetry, that magical beliefs – or the psychic mechanisms that support them – are *ineradicable in the human psyche*.

If the mentality of religion and pseudoscience is not just protest or massage but a search for alternative truths, it is, as we say, whistling in the dark, looking for meaning and subjectivity without a clue where to find it. Instead of proposing magical belief as an alternative or corrective to science, one should begin with the holistic forms that are its ancestral sources, not with the 'manifest content' that is an end-stage of analysis. The illogic of the primitive diverts us from what is authentic in it by the sheer multiplicity of its products. Magical thought, from taboo and religion to telepathy and clairvoyance, does not reveal the character of the subjective so much as its needs made concrete in a world of fact. In non-cognitive entities, the subjective is becoming and duration, i.e. relationality and temporal existence. In cognition, the diachronics of meaning-creation erupt as a self-expressive mode of archaic thought-process released from the constraints of objectivity.

For science to include the subjective in its scope, a shift in its methods and presuppositions would be necessary. Such a shift, were it to occur, should follow the historical pattern and arise *within* physical science itself, for science is not monolithic, and there are powerful controversies that might be expected to bring about radical change. Yet, the reaction to science does not seem a portent of a radical upheaval in physical theory that is just around the corner. And within science, the competing paradigms of relativity, chaos, quantum mechanics, the anthropic principle or superstring theory, have not been occasions for a global reformulation, for each theory is useful within its domain of application. Contradictions coexist, or are ignored, or are assimilated to current doctrine. The hegemonic tendency is there but it does not seem to apply to the contradictory impulses of current day physics. Instead, there is a pluralism of belief, even at the most fundamental levels.[8] The challenge is to find a deeper theory, not a new physics. That is why if physics is to account for consciousness and subjectivity, the solution, in my opinion, will *not* arise in physics, nor, certainly in a validation of the supernatural, but rather, and this is the thesis of this chapter, on the evidence of cognitive studies that a new understanding is required if we are to know the nature of scientific objects.

The ancestral line of human subjectivity

Subjectivism is the doctrine that the perspective of an observer is the primary source and substrate of knowledge of the world. Idealism is the outcome of a subjectivism that is rigorously conceived. From a microgenetic standpoint, idealism and its counterpart, realism, are centered at successive points in the *process* of object formation, idealism on the formative or psychic phases and the continuity of imagery with object perception, realism on the *conclusion* of the process in a concrete object with the self remaining behind as spectator. When the conceptual phases that generate an object are carried over into the object, the world becomes conceptual and subjectivism passes into idealism. When the concrete phases that deposit the object are projected back into the mind, the conceptual objectifies and realism passes into materialism.

Until the radical materialism of the present century, subjectivism had been the dominant theory of mind and nature. If we ask the materialist why this should have been so, we will probably hear that it is a vestige of primitive beliefs, in demons, souls or spirits, in effect a mythology purified and turned mystical. While the validity of subjectivism is not in danger from science – even a retreat of subjectivity into feeling, sense-data and value leaves, in my opinion, an ineluctable residue that will continue to defy quantitative measures – the materialist does have a point, namely that subjectivism is an outgrowth of the primitive carried into ordinary cognition. Indeed, there is a continuous thread of animism from an unreflective belief in the supernatural to the most rarified systems of philosophical thought.

Scientific objects are the culmination of historical process, but they emerge out of the primitive, as do all objects, in every act of cognition. The invisible ancestry of the mental state, not the lively contents of conscious thought, empowers action with agency and belief. The generality of animism in a diversity of cultures discloses, at its very beginnings, patterns in magical thinking that are basic to human cognition. Durckheim and Mauss, and Levy-Bruhl[9] described an initial stage of diffuse subjectivity continuous from inner to outer, pervasive like the Force in *Star Wars*, not an external power in confrontation with rationality, rather a realm of magic continuous from mind to world. An individual arises in a pool of spirit-nature, not as a spectator but as a goal. Gradually, the inner and outer segments of the process are allocated to different portions. Eventually, the process cleaves to a self-identical subject and a mind-independent world. The individuation continues to populate the world with objects and punctuate the mind with entities. The specification of object-concepts accompanies the individuation of a person in seeming opposition to the rest of nature. Then, abstract concepts differentiate an interior self.

In the course of this process, an object concept arises in a field that includes the world, then sequesters in the mind where it originates. The attribution of mind to nature is prior to the experience of mind as intrapsychic. First the mental is apprehended outside the mind, then intrapsychic phases undergo a development within the observer.[10] The subjective is first situated in external objects, such as the totem of a tree. Almost every object has a psychic or sacred quality. In the growth of the mind, the conceptual investment in devotional objects remains behind as the meaning of the object in the mind of the observer. The conceptual feeling in primitive objects is the basis of value in mature ones. This 'projection' of mind into nature is not a relocation of images from the mind to the outside world, but a progression from the intrapsychic to the extrapersonal, which is the direction of thought. The mind does not fling itself outward into a world that is waiting to receive it but actualizes the world as a distal segment.[11]

With increasing abstraction, the spirit of the tree becomes a god of the woods. The category is invested with the meaning formerly concentrated in a member, while the particular retains a portion of the psychic as it generalizes to the class. Intermediate phases can be identified, for example, the transition from the theriomorphic representation of animal spirits to the anthropomorphic representation of gods, the animals persisting as their companions, such as the lion of Shiva or the owl of Athene. The abstraction accompanies a specification of object concepts and an extraction of the subjective from the objects of experience. The god of the woods becomes a deity of nature, more generalized than a totem, more external, inclusive, less concrete, largely independent of the observer. There is a mitigation of power. Magic compels, religion persuades.

A progression from myth, to religion, to science is thematic in accounts of the history of thought. In this transition, each new stage can be conceived as a derivation of the prior one. Frazer (1927) wrote that the anthropology of rudimentary societies was the study of the 'embryology of human thought and institutions'. Freud recognized a shift from myth to science in cognitive maturation, from narcissism in magic, through object choice in religion, e.g. the father/child relation in the attachment to parents, to adaptation in an object-centered world. The stages are not bypassed and forgotten, but persist as subordinate modes or subsurface complexes that motivate and configure the contents of consciousness and behavior. These constructs account for the tension between a world of adaptation and an equally real universe of psychic phenomena; Frazer's roads of outward and inward experience, Levy-Bruhl's worlds of the everyday and the unseen.

One can say, two systems of belief actualize in every cognition, the belief in the reality of the world, and the reality of the world of belief. A myth is a reminder that our dreams are also worlds. As dreams are the forerunners of momentary thoughts, so magical thinking is buried in every act of cognition. Whitehead wrote, 'a myth satisfies the demands of incipient rationality'.[12] The supernatural is very much a mother to the child of reason, in the words of Coleridge,

Til Superstition with unconscious hand
Seat Reason on her throne.

The conceptual nature preserved in mysticism is a background of feeling that permeates the surface detail. The mystical is the intuition of a deeper purpose, what Wordsworth referred to as the 'gentle agency of nature objects'. In mystical thought, a growing abstractness strips the external of human attributes and veers inward in surrender to the homogeneity of one's own subjectivity. Mysticism is the beginnings of philosophical reflection. The purpose of philosophy, Whitehead said, is to rationalize mysticism.

The doctrine that the phenomena of nature reveal the presence of an immaterial spirit is a mode of thought that duplicates in the external world the subjectivity of the observer by infusing into nature some attributes of the human imagination, such as agency, will, purposefulness and awareness. Though a topic of dispute, Lévy-Bruhl argued that the primitive mind has a rudimentary self-awareness and sense of personal existence, not a self-concept in the sense of an identity that stands behind and scrutinizes states of agentive feeling. The evidence of neuropsychology is that the relation between a self and an object, or a proposition, cannot be postulated as a ground for belief or agency. The distinction of a self and its mental or perceptual contents is an achievement of process and a goal of thought. Introspection requires an attenuation of preterminal phases to reclaim objects as inner experience. When the phase-sequence does not arouse an introspective preamble, there is no decision, choice, awareness of self or privacy.

The animist does not *see* a different world, nor does he apprehend the conceptual as a projection. He *feels* in the object a penumbra of meaning, just as the psychotic attributes a thought-creation to an external source. Feeling is prior to knowing and the basis of conviction. Neither the primitive nor the psychotic can be convinced of the illogic of their beliefs. Every belief takes experience as its material, and is a mix of the real and the imaginary, the known and the unknown. If a primitive mentality overflows in a modern setting, the individual will be diagnosed as a psychotic. A villager who believes incontrovertibly that his brother is a crocodile is at home in the jungle but delusional in New York. We are close to Nietzsche's definition of a belief as an irrefutable error.

The nature of primitive mentality

For the most part, anthropology in the late nineteenth and early twentieth century regarded the primitive in thought or culture as the mark of a child-like, rudimentary or prelogical stage in the growth of the mind. Subsequently, an awareness of the richness of primitive language and society, the interpenetration of magical and scientific thinking, or the blending of imagination with adaptation, the dearth of universals in a diversity of cultures, ancillary factors such as illiteracy, climate, tradition, and especially, the advent of cultural relativity, marginalized the significance for philosophy of mind of the work on primitive mentation. As a result, myth, magic and the notion of the 'primitive' came to be seen as the products of an attempt by guessing to fill in gaps in prescientific knowledge even if, on these grounds, the cross-cultural similarities could not be explained.[13]

But myth as an explanation of the unknown is not mere supposition. The internal coherence and similarity across cultures suggest the influence of common 'schemas', procedures and categories. If magic and myth were fabrications strung into narratives, accounts of the unknowable would follow the laws of causal science, even if the explanations were vacuous. The needs of conceptual assimilation compete with those of perceptual adaptation. Why would supposition postulate forces incompatible with perceptual experience? Why would human needs and lines of expression deposit culture-specific entities independent of factual encounters? For Frazer, magic was the result of primitive associations, coexisting with rational behavior and appearing where knowledge is thin. I would say the 'associations' of magic involve a paralogic that foreshadows those of reason, which surface when conceptual feeling falls short of full analysis.

The relegation of magical thinking to guesswork without a deeper significance for theory of mind resembles the dismissal of dream content as neural firings in the brainstem, with meaning applied on waking to images that are conceived as random noise.[14] In fact, the search for cognitive invariants in myth[15] suffered the same fate as dream analysis, i.e. irrelevance to philosophy of mind, to which myths are often compared. A myth, so the analysis goes, is a guess about unknown aspects of nature, birth, sexuality, illness or death, that are mysterious, terrifying and unpredictable. The skill, ingenuity and imagination the analyst brings to a set of general categories, in dream or mythic *content*, appeals to no standard other than plausibility. Since the interpretation must accommodate conscious *and* unconscious thought, the real *and* the psychic, the value of the content and the validity of the interpretation depend on the ability of the interpreter-sage to elaborate a *container-myth* that exceeds the original in its explanatory power. When we analyse a myth we give an interpretation that is rational to one that is unconscious. Often, we seem to be asked to exchange the impossible for the merely implausible.[16]

If the analytic products of an act of thought are conceived as free-standing, and if a rational consciousness is conceived as equivalent to thinking, i.e. if thought is restricted to the 'interactions' of conscious contents in an abstract logic space, there is little room for myth, dream, pathology, affect, instinct and the unconscious. These phases are excluded by a mandate that limits inquiry to contents the histories of which are dimly understood and of less than perfunctory interest. To go from culture to mind requires a theory of cognition to guide the study of culture, and such a theory cannot afford to ignore the antecedents and the context of the logical in pursuit of a model of the rational. Reason is paralogical before it is logical. This too has a part in theory. We embrace, even celebrate, an identification of the creative with the archaic, with dream or with madness. Like dream in relation to waking cognition, the primitive is the secret heart of a conceptual nature that pulses into a world of adaptation. The archaic is not displaced but submerged. The past may be hidden but is never lost.

Paralogic

Magical thought unconscious of its origins manifests them in its occasions. The irrational, the affective and the symbolic (Lévy-Bruhl's 'affective category of the supernatural') refer to holistic, pre-analytic phases antecedent to discriminative or scientific concepts.[17] The terms syncretic and paralogical, the mechanisms of fusion, symbol formation and condensation, that are found in dream, trance[18] and psychosis, are descriptions of this mode of thought. The objectified thoughts of the psychotic are equivalent to the sacred objects of the primitive. False beliefs acknowledged as illogical and held with deep conviction are not emotive or neurotic capsules. The sense of realness, the certainty (lack of doubt) and the affective strength of dream, magic, myth and pathological cognition point to phases prior to the elicitation of analytic concepts in an intrapsychic field of choice. The primitivity in mind is the preliminary in the mental state. Magic and dream discharge antecedent phases in the actualization of a thought. These phases are traversed in every cognition. We fall in love, burst in tears, have novel thoughts and irrational impulses, for reasons we do not understand. Being moderns, we reject the notion of a primitive cognition, or think those quaint who embrace it, and postulate causes in a forgotten childhood or a buried unconscious.

Freud interpreted animism as narcissism with an overvaluation of the psychic (incidentally, a good description of psychoanalysis), where mental components are locked like gods in mortal combat. Psychoanalysis takes the products of mental activity as explanatory forces and replaces the laws of actualization with an interaction of entities at imaginary surfaces. My goal is not to depict the primitive as a form of pathology, but to integrate the pathological and the primitive with a theory of normal cognition. Paralogic is the mechanism that these forms of irrational thinking share. In paralogic, topics are identified based on shared predicates. There is an inability to distinguish resemblance from identity. We see this, normally, in frozen metaphors – *the brain* is *a computer* – or in dream symbols, e.g. the substitution of knife for penis.

Paralogic is the identification of objects through an overlap of attributes that are not just connecting links but have a certain parity with the entities that share them. The concreteness of primitive thinking is largely a result of the heightened reality, or objectification, of attributes, and the intrinsic relatedness of properties in the objects they modify. An objectified attribute is a whole in relation to the entity of which it formerly was a part. In paralogic, wholes and parts, or the relation of subject/topic to property/attribute, are not in an abstract relation to each other, such that the part symbolizes, represents or *stands for* the whole, which is an abstract version of a concrete situation, rather, the parts *participate* in the whole through a kind of conceptual or affective fusion. The part *is* the whole in a real, concrete sense. The relation is not one of a part as a feature or a whole as a sum. The whole is in the part, and as real as the part is.

An example of this in primitive cultures is the sense of a community (whole) as an entity, not a collection of individuals (parts). There is a real and powerfully felt sense that the cult is prior to the personality and that the group is wounded if one of its members is harmed. Belonging to the group is a primary fact of mythic awareness. The group or clan extends into past and future time, and includes living, dead and totemic ancestors. The individual is so much a member of the group that death may be apprehended more as a loss of membership than of individual life. This primitive belief may yet survive in the sense that we all have, to a greater or lesser degree, of empathy for others and a shared humanity.

There are other renditions of this theme, for example, the extension from the individual to the community, as when the killing of a snake is forbidden for fear of retribution from the population of snakes. This is a generalization from the particular to the class, or a return to the category out of which the particular individuated. Lévy-Bruhl claimed the class replaced the individual as a concrete entity. A class is abstract in relation to the particular when the relation of whole to particular is implicit, either in the particular or in the whole, but the class becomes concrete when, for the sake of the whole, the particular is suppressed. The felt concreteness of paralogic extends to phenomena such as metaphor, which is a sorting of types by a partial similarity, often with a quota of feeling or valuation. In creative thought, metaphor is a bridge to a novel perspective. The imagery, connotation and creative potential of metaphor underlie and give rise to the rigor of mathematical logic. Metaphoric images accessed in a mature cognition, where the archaic coexists with the recent, and the recent is dominant, represent a phase beyond paralogical fusions in which the archaic mode predominates. Logical propositions articulate, expand and individuate the relations of subject and predicate that are compressed in the fluid coloration of paralogic. In effect, subject/predicate logic develops out of predication.

More precisely, logical thought recognizes a similarity of properties between objects as a resemblance of the objects in respect of those properties, whereas the similarity of the objects in primitive thought (or in cognitive disorders[19]) is sufficient for their fusion. A man who is like a leopard in certain attributes *is* a leopard. He is invested with leopard powers. The paralogical syllogism is of the form: Akira is swift/leopards are swift/Akira is a leopard.[20] Shared attributes signify identities, not partial relations, whether accidental or necessary. For the Bororo of Brazil, the Trumai are men who sleep under water.[21] Such beings are only possible if they share properties with fish or belong to the same category. In dyirbal, a language of the Australian aborigine, of the four primary categories, men and fish are in the same category of *animate* things.[22]

Belief and knowledge

Paralogic generates propositions, logic decides if they are true. Logic can prove to be false what we feel to be true, but often we cling to a false belief and only later, reluctantly, feel the truth of the falsification. Paralogical constructions are variably shaped by the demands of logic. The one elaborates, the other edits. But logic is often powerless against the intuitive 'truth' of primitive ideation. The products of paralogic are *felt* to be true, logic has to convince.[23] The schizophrenic feels an idealist philosophy the idealist only talks about. Knowing we will die we act as if – thus believe? – we will live forever. Is denial of the obvious a repression of the painful or is it, in the will to live and its ghostly companion, the belief in the soul, a conceptual primitive surging into the present occasioned by an instinct for self-preservation? If the self and world were decisively proved to be phenomenal, how many other than bodhisattvas would believe (feel) it? Is it not as much a question of how real the object *feels*, its palpability or the (Platonic) grounding of objectivity in presentness, than a compilation of evidence for or against its truth?

The antecedent in cognition tends to be accompanied by certainty, the consequent requires (seeks, deposits) proof. The potential of antecedents delivers paralogical objects into a primitive cognition and options into a mature one. Whatever is felt as real cannot be disconfirmed. If there are no choices, the choice of a real or unreal (imaginary) situation does not arise. The feeling of realness may be more pronounced in those with little gift for introspection which would otherwise permit options that undermine certainty. Options entail doubt which, as Descartes famously argued, is the beginning of reason.

The antecedent is closer to fantasy or 'primary process' thinking, the consequent is 'tested by reality' or adapted to the world. The affective element in deeply held beliefs, whether implicit or conscious, derives from the strenght of antecedent phases. The more archaic, the more intense the feeling tone, the greater the affective tonality of the developing content. The generation of conviction by feeling reflects the permeation of belief with certainty, whereas the waning of affect at the cognitive surface, as valuation distributes into objects, transfers feeling into the final contents of cognition, thus depriving the antecedent belief of its emotive power.

Belief doesn't become knowing just because there are grounds for certitude, for certainty in knowledge may be prior to the provisional in belief. The surfacing of presuppositions in the form of knowledge or explicit belief can make one conscious of the strength of one's certainty or doubt, yet we are guided in our actions by presuppositions that are largely unconscious. There are beliefs to which we are relatively indifferent, that are more like opinions or preferences, there are beliefs we will die for, and beliefs, largely implicit, that maintain sanity, such as the belief in a self or an independent world. The certainty in an unjustified belief is supplemented by a conviction that is more like a faith. On the other hand, when knowledge falsifies belief, the result is a new belief but one that may lack the 'animal faith' of the old one. I know she loves me, I believe she loves me, I know she does not love me. Joy, confusion, sorrow, as belief (and personality) undergo a transformation from one extreme to another.

The conviction that accompanies paralogical belief is nearly unshakeable. To be unable to convince someone that men are not crocodiles, or to persuade a 'true believer' that Martians have not landed on the earth, has astonished all who have conversed with such people. We wonder at their obtuseness or intransigence. The same certainty is found in delusion in psychotics and in individuals with brain damage, who give erroneous responses with confidence and rarely say they do not know the answer. In such cases, the penetration of belief by conviction is unrelated to evidence. The sensitivity to evidence supposes a mode of thought adequate to perceive the 'real', thus a nearness to facts that is of a piece with the attainment of reason. An openness to reason is the mark of a rational cognition. To employ reason to refute paralogic is to oppose a mature cognition with a phase it has already negotiated, one that actualizes in the psychic objects of the primitive. Like a piece of driftwood that is carried downstream, a primitive belief is a distal outpost of an unconscious operation that occurs early in the trajectory of a thought.

Conviction in a belief depends on the realness of its contents. To believe a man is a crocodile is to feel the realness of this belief in its objects. Realness is linked to an immediate presence of data in consciousness. A belief is a psychic entity, but the feeling of reality or certainty requires a concrete situation. The belief must be fulfilled by actual events. The reality of myth is the realness of its constituents, which are not mere thoughts or images but concrete happenings. In effect, the feeling of reality is the objectification of a belief as an actual world.

In primitive mentality, myth and dream have an immediacy that is concrete and real. Lévy-Bruhl wrote of the aborigine, 'the world into which their dreams lead them is hardly, if at all, to be distinguished from the world of the mythic past'.[24] The aboriginal term for dream is the same as that for myth.[25] Schizophrenics also experience a dream as an actual happening. Some aspects of their waking life have the character of dream cognition. Like a myth, a dream is not a thought that comes and goes, but an actual event that has an impact beyond the state of its occurrence. (Is writing not an attempt to create enduring (sacred) objects out of transient thoughts?)

A psychotic delusion is comparable to a personal mythology – a deeply-held illogical belief – that acquires an irrefragable force. Like a myth or totemic belief, delusion infiltrates psychic objects yet coexists with abstraction for events unrelated to the delusional content. The schizophrenic believes he is tapping into other people's thoughts, delusions and dreams. There is a two-way traffic; other people tap into his mind as well. The control of his mind by an external agent resembles the influence on the primitive of a totemic object. The psychic objects of the primitive and the schizophrenic tend to be persecutory.[26] There are similarities in the onset of hallucination, a stage that reflects the incomplete actualization of an object or, what amounts to the same thing, the retention in the object of intrapsychic qualities. The schizophrenic hears voices in the rustle of the leaves. The Maori described by Lévy-Bruhl hear 'voices of unseen things in the rustling branches, in whispering winds, in the sound of rushing waters'.

Primitivity and cyclical return

Ancestral patterns illustrating mechanisms of paralogic survive as encapsulated fragments that influence behavior even as they maintain their resistance to logical scrutiny. The abuse and sexual mistreatment of children by parents or guardians awakens a powerful sense of abhorrence that receives a quotient of its emotional force from the incest taboo in which it is grounded. The elicitation of memories of incest in the course of psychotherapy, whether true or false, revives in the passivity, suggestibility or semi-trance of the therapeutic state, an ancient mentality still alive in modern thought. The prohibition of incest has an emotional intensity and structural complexity that, as Freud argued, identifies a taboo in the suppression of the forbidden. The forbidding is also an inducement. The taboo can be a bridge or a barrier, just as a property or a feature can be a path to another object or a boundary, but in either case it is an affectively charged situation that is balanced on the relation of part to whole, child to parent, individual to clan.

Incest taboos are shared with tribal people when clan relations supersede blood ties. This is an instance where the group – as a psychic whole, not a mere aggregate – precedes the individual. The notion that the incest taboo developed as a procedure for male bonding through the exchange of women as property is consistent with the fact that the taboo appeared in tribes prior to a concept of male paternity, that it may have begun with a prohibition on the union of brothers and sisters to permit the exchange of sisters for other women or equivalent property, that it prohibits unions within clans that are not by our standards incestuous but permits others across clans that are, for example, a father–daughter union permitted in a clan with a matrilineal descent.[27] Tradition and the lineage of ancestral spirits are more potent than paternity in the life of the individual. In modern societies where the parent/child contact is primary, the taboo is as strongly enforced, even when there are no risks or blood ties attendant to the union, such as that of father and stepdaughter, or parent and adopted child. Yet we do not think the incest taboo would be less pronounced if science were indifferent to the genetic or psychological consequences.

Remnants of primitive mentality exhibiting the part-whole relation abound in developed cultures. The sophistication of my French wife, Carine, does not prevent her from continuing the tradition of her Algerian mother in placing a knife under the mattress to ward off devils.[28] Here, the dangerous properties (functional parts or attributes) of a knife serve to connect or separate an individual and an evil spirit. An elderly Italian woman of my acquaintance, of royal blood, a professor of philosophy, sets a place each day at the table to converse with her long dead husband.[29] This 'conceptual' synechdoche, in which the part substitutes for the whole, *pars pro toto*, satisfies a primitive need that is incompatible with logic. A Swedish professor of neurology performed a bris on my son, Ilya, and then followed the orthodox custom, again, a part for the whole, of burying the foreskin with others he has collected which, when he dies, will be placed in a common grave. When questioned on the logic of this action in relation to cutting the hair or fingernails or the elimination of body waste, he gave an emotional defense of the practise with obvious conviction.[30] Examples in which magical belief coexists with rational thought at the highest level can be multiplied by the thousands.[31]

Science and superstition

Magical beliefs that lie outside the parameters of science are accorded a skepticism beyond that ordinarily reserved for the merely unproven. In modern life, when the potential of the unknown is too hastily banalized by an example that is excessively concrete, primitive belief turns a mystery into an object of ridicule. Scientists, however, will often tolerate a bit of mysticism when it falls beyond the scope of method, for example, in the concept of deity, or in the profound ambiguities that bedevil epistemic inquiry. The weight of interest in those 'primitive' ideas that still flourish in contemporary science is proportional to the latitude over which the subjective is condoned.

Consider the mind-dependence of time in relation to the dreamtime of myth. Lévy-Bruhl wrote, 'myths do not unfold in time; on the contrary, it is time which is in the myth'.[32] Is time discovered or created, real or illusory? Are we to follow Russell or McTaggert? The notion of a *now* on which the past and future converge – taken seriously in some isotropic theories – has been debated in process metaphysics, and is central to some schools of

Buddhist philosophy, e.g. *Hua-Yen*. The simultaneity of the dreamtime is prior to the seriality and incrementation of wakefulness, as the logical and analytic are consequent to the syncretic and holistic, in the evolution of mind and culture, and in the momentary history of a thought.

Mutations of shape and identity in magical objects are related to the concept of objects as accidental forms of change. The stability of an object is realized out of an antecedent plasticity which is the potential in a configuration for a multiplicity of realizations. The form and stability of objects are not given but achieved. In hallucination or illusion, a distortion is a sign of incompleteness. In positivist theory, the realization of an object is presumed to be accomplished through a constructive process beginning with sense-data, but it could also occur as a graded elimination of non-adaptive possibilities. The ubiquity of the psychic in the world of the primitive conforms to an idealist theory of objects as objectified ideas.

In idealism, the presumption of an equivalence of image and object in the facticity of experience, i.e. the imaginal basis of objects or the conceptuality of perceptual nature, is the legacy of a primitive mentality submerged in a mature system of concepts. The spacetime of magical thinking – the possibility of action among entities separated in space and time, or the image of two entities that share the same location – follows patterns of precedence, not causal interaction. While a non-causal theory of nature is alien to scientific thinking, it is consistent with accounts of *spooky* events at the quantum level.[33] Of course, even for the physicist or the philosopher, a belief in nature-spirits and non-causal change survives in the everyday concepts of luck or just desert, or when we give thanks for an unexpected blessing, or feel in our heart that a misfortune is a punishment.[34]

In magic and superstition, the mind acts on objects, external agencies act upon the mind. Spirits are the intentional agents of an occult nature. The spirit world of nature resembles a mind in having intentional attitudes, but differs in the entities that exercise them. The properties of cognition are exchanged with those of nature in the idea that the mental state is a set of causal links with personal choice determined or fated, while only God has agency. The transfer of human freedom to the will of God is a reversal of the norm in experience, that choice and decision are expressions of human thought. Cognition becomes machine-like, nature becomes cognitive. This reversal, a common aspect of religious and philosophical thought, may represent a fragment of the phase of exteriority before the mental is experienced as intrapsychic.

The infiltration of nature by thought reinforces the suspicion that the laws of mind, inaccessible to introspection and yet to be clearly articulated, condition the laws of nature, or the reverse, or that both supervene a common process. Most scientists do not assign mental properties to nature, in crocodile gods or western divinities, but do locate mental faculties in the brain, which is a more likely candidate for them than a crocodile, but still reflects a concept of localization that is not inherently different. The localization of functional entities in the brain is motivated by the same pattern of thought as the localization of cognitive entities in nature. Localization extends to the psychology of the brain the one-way direction of whole-part analysis in nature and cognition.[35] In the absence of an account of covert process, the attributes of nature's mind, no less than those of the observer, are little more than the conventional icons of an *ad hoc* faith.

Introspection

The terminus of an act of cognition is an object in perception or an actualization of will in the direction of the body or an object. Normally, in perception, an object is fully exteriorized, while in action, agency is exerted primarily in the body, and secondarily, through the body, on objects. Agentive feeling 'belongs' to the subject or the body in a way that objects do not, retaining the subjectivity that objects have lost in their outward migration. The object achieves independence, while the action is kept within the body, carried outward in the direction of the perception from the depth of cognition to its implementation in thought or action. The terminus of this development, in limb movement or vocalization, embodies the experience of agency as a conduit from mind to nature. Agentive feeling is felt to originate in the mind as the decisive factor in action, but it can extend to nature as well, for example, when human-like agency is attributed to a nature god or divinity.[36]

The machine view of nature and the zombie view of man pervasive in present-day science have resulted in a near elimination of agency in the world and in the mind. But when we look back at the historical growth of the mind, or if we thoughtfully examine our own mental states, we understand that the agency of magical objects is not a projection or an inference based on fear or ignorance, but is a property of the mind in which they are elaborated. This mentality deposits in objects which are, ultimately, psychic entities with all the properties of the mentality that shapes them. The primitive object receives the conceptual feeling that normally exteriorizes in the world, and it receives the agentive feeling that normally deposits in the body.

Activities of the mind that are more like powers than objects discharge intrasubjectively. Feelings and intentions flow into the truncated (mental) objects of primitive cognition and deposit as the emotional tonality of demons, e.g. kindness, cruelty or vindictiveness, and the willfulness of playful spirits. With the appearance of introspection, the interior objects of the primitive that discharge in perception now emerge as the conceptual precursors of objects that are fully exterior. The subjective phase that exteriorized as psychic content undergoes an expansion prior to the commitment to a specific object. The priority is the potential for a multiplicity of choices antecedent to the one choice that actualizes.[37] A veridical object is sculpted to actuality, but its potential remains behind as introspective content. A suspension of subjective aim accompanies a shift in will to agency and the double nature of objects as perceptions and ideas.

The primitive mode engages the subjective at a point in the evolution of mind and culture when objects have psychic qualities but are not yet entities for introspection. These qualities are carried through into objects as they exteriorize. Primitive thought and phenomena such as dream mentation and some forms of psychopathology, confirm microgenetic concepts of the subjective as referring to that segment in the actualization of the mind/brain state that lays down the contents of privacy en route to an actual world, an actualization that is categoreal and wavelike, segmented into phases that are regarded as ontologically distinct, as in the passage from the physical brain or unconscious to the conscious and subjective, or from the subjective to physical objects. In this transition, every occasion of knowledge involves a restructuring of thought and, at each moment in the revision, opportunities arise for the prominence of one or another segment, for example, a state of reflection giving way to one of object awareness. The dynamic of the phases and their disproportionate influence in an act of cognition determine whether the cognition is closer to concepts and meanings, or objects and words, closer to the primitive or the analytic, to the creative or the habitual.

In the growth of scientific thought, novelty is replaced by inevitability as primitive 'associations' give way to causal relations, but the primitive persists as a foundation out of which causality develops, with the psychic surviving as the power in causal necessity. Causal theory requires for its support a trace of the magical that reason has all but distilled from primitive objects. That is, the power of a cause to carry into (induce, transmit) an effect – the 'psychic addition' to causal nature – is the residual agency of the psychic in an exteriorized *mental* object.

We can speculate on the origins of introspection from the double aspect of animistic objects, in which an entity is both substance and spirit, attributed by Freud to the persistence of the memory of the thing after the perception has ceased. The capacity to revive an image of an extinct object is an essential feature of introspection. If Freud is correct in this analysis, as I believe, in part, he is, the double aspect might provide evidence of the capacity for introspection in primitives, even if it is weakly developed. Conceivably, the initial step in the emergence of introspection is the arousal of a memory image, which is as real to the observer as the object to which it refers. The potentiality and omnipotence of the imagination take on the attributes of concrete powers attributed to an external spirit. A stage of concrete imagery is followed by a recognition that the image can be detached from the object, compared and contrasted and brought into relation with past and future events, in other words, freely manipulated in thought.

Belief and fact

The resistance of magical belief to empirical fact raises questions of a more general nature. Thought is a mix of the available and the unexpected, the latter being a mark of novelty in relation to a goal of adaptation. What finally materializes is what remains after the concessions to novelty have been effected. A belief corresponds to what a fact points to, in addition to the unexpressed residual after the fact is subtracted. It could not correspond to the fact itself, which is an instance of the belief. Facts are creations, not encounters. There is only a kind of adequacy in the degree to which a belief is realized, or the degree to which a fact satisfies a belief. On the other side of the fact is its necessity, i.e. the extent to which error is shaved on the way to ecological fitness. The realness of the fact from a non-cognitive standpoint is just this fitness to noumenal events.

True and false beliefs have much the same cognitive history but the former are felt to be the outcome of observation and consensus, while the latter are conceived as free inventions. True beliefs are object-centered and adaptive, false ones are not, even if they form a relatively coherent system. A false belief that is unconstrained by experience is, by definition, not analyzed to a scientific concept. A false belief, whether in superstition or in science, discharges in fantasy, half real, half imaginary, at times to the point where, if it is powerful and persistent, it is labeled as a delusion. A scientific delusion is a serious belief posing as truth that consumes the interests of an individual or is endorsed by the scientific community. The passage from a belief to a fact or to a delusion is a function of its testing by experience, which is also a sign of how far the belief has objectified. The fluctuation from belief to fact and delusion is a reminder that there are no mind-independent facts. Regardless of whether a fact has support in opinion or experience, a belief is lurking in the background of every fact. This is not just because the fact ultimately devolves from thought, or that all documentation is theory-driven, but because a fact is not an intrinsic reality that is independent of knowing it, it is the momentary objectification of a belief.

Facts justify beliefs – the whiteness of snow justifies the belief that snow is white. This is to say that a belief (about snow) realizes a fact (white snow) that is judged to be adequate, in this case by others who have the same belief that is fulfilled in similar perceptions. The whiteness of snow has to be confirmed by other commentaries. The roundness of an apple is confirmed by the additional modality of touch. An apple that looks round and is felt like a cube might be judged an hallucination in vision or in touch. The belief that apples are round might justify vision, or disconfirm touch. If the confusion resulted from a brain disorder, one could not say the belief that apples are round would override the evidence of touch. The more 'palpable' of the two realities might prevail. The fact is both an instance of the belief and a sample of its validity. This is confirmed by studies in neuropsychology. For example, the disbelief in the existence of a hand with a loss of tactile feeling is not reversed by normal vision. Conversely, when the feeling of reality in a phantom hand is challenged, a tension arises between the visual and somaesthetic facts in relation to prior and occurrent beliefs. The loss of (proximal or distal) feeling can occasion a belief that a present limb is absent or that an absent limb is present.

That a correspondence of a belief with a fact can be overridden by a contrary fact implies that beliefs are hostage to facts, or that they do not compel the facts that instance them. I think this means that a belief is the potential to be known by a fact, while a fact exhibits a belief in its potential to actualize into what it is by way of constraints imposed on the actualization process. In everyday situations, a belief may be resistant to a contrary fact because the belief is generating the facts, and in so doing, disclosing what is possible in its potential that otherwise would not come into being, i.e. the facts comprise whatever portion of the belief they realize. Those 'portions' not realized abide in the potential (or disposition) for another actualization. In brain disorders, the erosion of a belief when a fact is altered is a sign that the belief does not exist other than as occurrent fact, i.e. there is no entity of belief with phenomenal properties apart from the contents that are realized, and the constraints that determine what those realizations will be.

Subjectivity and philosophy of mind

Animism is the primitive in subjectivism, as a machine theory of mind and nature is the primitive in materialism. An animism that restricts the scope of the subjective, or a materialism that admits an irreducible subjectivity, is close to a dualism in which mind and world are distinct. Where there is an inexplicable gap between the mental and the physical, dualism creeps in. A deeper understanding of the nature of subjectivity, and the reverse, a theory on the subjective in nature, will avoid an outcome in dualism, which is inevitable with a disjunct of the conscious and the non-cognitive, i.e. if the subjective is equated with consciousness or is not interpreted on a continuum of physical evolution. The presumption of a non-conceptual boundary is crucial to the role of a material substrate. If the material achieves a complexity that is sufficient to generate a state of consciousness, and the transition is continuous, rudiments of consciousness would extend all the way down and the system is panpsychic. If the transition is abrupt or emergent, and the emergence does not have a physicalist explanation, the system is dualist.

The inability to link the qualitative to the quantitative, mind to brain, or the conceptual to the physical, gives rise to the doctrine that it's all just physical, and that qualitative feelings are nothing more than the functional states of a physical system. What is left over in this assumption is the translation from brain to mind state, i.e. how we derive cognitions from neurons, and the phenomenal quality of the mental. In contrast, the uncoupling of mind from its moorings to the physical, with access to the subjective but not the physical component of the state, gives rise to the assumption that it's all just mental, or that the physical is an inference about the sources of the mental, which is an inference about a non-cognitive past from an experiential present.

It has to be conceded that a subjectivism that ignores the world is as bankrupt as a materialism that eliminates the mind. Subjectivism should not have to choose between a reality that is beyond the grasp of the conceptual and one that is subjective to its limits, and materialism should not be forced to a simplification that defies credulity or a dismissal of what it cannot explain. One can postulate that mind is everywhere, or that mind is nowhere, but to argue that mind is *somewhere*, and that wherever, or whenever, it may be, it is independent of the physical, implies that nature is one part conceptual and one part material. (These irreconcilables recall the consubstantiality of totemic belief, where a psychic image and a physical object are conceived as double existents.) An acceptance of this dual state of affairs is a compromise of the unresolved tension of subjectivity in relation to brain activity. It has all the drawbacks of a pure subjectivism without the coherence of idealism. From the standpoint of the motivational roots of metaphysical suppositions, dualism thrives on the seeming recalcitrance of mental and physical properties, i.e. the incompatibility of assertions as to their nature, fueled by an emotional response to the dissolution of material existents and the inevitable annihilation of consciousness.

A reduction and/or elimination of conceptuality is the preferred strategy of materialists. Such a reduction is possible only with an objectification of process and a neglect of the temporal dynamic of brain and cognition. Efforts to accommodate mind and brain in a relation of causality or identity without eliminating the conceptual reconcile the mental and the physical by postulating rules of correspondence. Such rules imply a translation across vocabularies. But, if similar laws apply to the mental and the material, bridging principles are unnecessary.

Some philosophers allocate the conceptual to what it's like to be some sort of creature, and its byproduct in *qualia*. The gulf that separates a qualitative feel from the methods of objective science is a wedge in the continuity of process. I wouldn't know what it's like to be a bat, or another person, but what could I say of what it's like to be myself, awake, drowsy, angry, joyful, dreaming? The question of what it's like to be another person does not have an answer. An imponderable is not an argument. One cannot draw a substantive conclusion from a statement in which ignorance is used as a key premiss.[38] The feel of a subjective occasion, to my way of thinking, is a non-starter in science and a reactionary position for philosophy. The experience of pain, or the *quale* of redness, is a fashionable testing ground for the warring camps, but the defense of qualia seems already a retreat. The world itself should be the field of battle.

The difficulties with a subjectivism that replaces physical nature with a universal conceptuality, or one that admits a physical substrate for some part or all of the conceptual, have forced many scientists to reject subjectivism, even in its most limited applications. If the dualist has a responsibility to explain the immateriality of the mental, the physicalist must account for its occurrence. The sources of mystical belief or religious feeling that lie in the interior process of organism are an antidote to this line of thought. Patterns of cognitive descent are the starting point for a survey of material objects as exemplifications of a common design. This commonality rests in the material process of the subjective, the becoming-actual of an object, or its interior life. A common process objectifies the psychic in nature and nature in the mind. The multiplicity of mental and natural objects owes to the becoming of wholes into parts where the wholes are the potential of categories and the parts are their incipient members. We, too, are wholes to myriad elements, and parts to a larger whole. From time to time we sense this contact, for example, in an experience of the sublime, when a portion of nature that is the mind of an individual dips into the pool of all-inclusive nature. This contact is nature's embrace of the observer. It is to be valued and contemplated. In this experience, the mind of the observer enters a dynamic hidden by everyday appearances to apprehend in direct communion, not through deliberation, which is a distancing from the world, or in daily intercourse, which is a partaking of the world, but in a sudden illumination, as a blessing.

Chapter 9
On Aesthetic Perception

We have all had the experience of viewing a painting unprepared and, over time, especially with explication, gradually coming to see the work in a wholly different light. We speak of a growing familiarity with the work, or an expanding store of knowledge in relation to which the work can be evaluated, but what does this signify for the original content? Is knowledge available to a bare object as an external resource – the library metaphor of common sense – or, however counter–intuitive, is it more deeply invested in the act of perceiving? Is aesthetic experience a special case of ordinary perception or a different mode of cognition? Familiarity and knowledge increase enjoyment of aesthetic objects, if not ordinary ones, but they do not provide an explanation for the process(es) they refer to; they do not help us decide whether the concepts that are emphatic in aesthetic perception emerge in the process of perceiving, or are juxtaposed to perceptions as extrinsic repositories, inculcated, stored in the mind and looked up when needed.

The problem has been forcefully put by Danto. He argues that content alone cannot distinguish an artifact from a work of art, for the latter is 'like an externalization of the artist's consciousness, as if we could see his way of seeing and not merely what he saw.' In this way, a work of art is said to differ from a 'mere representation', for it expresses something beyond the content of what is represented.[1] This something includes the feelings, the intentions and the conceptual sources of the work which are, in some sense, if not in the immediate content of the perception, in the wider perceptual experience, though in exactly what sense is not clear.

Art is distinguished from ordinary perception by its intensity of conceptual feeling. The passage from, say, an ordinary chair to one that evokes memories, associations and ideas, to a chair as an artistic or architectural topic, where habit gives way to novelty, or the innovative in thought replaces the replicative in memory, owes to the variation in conceptual feeling out of which these objects develop. But to write, as Danto does, of a *mere* representation, or *merely* what the artist sees, is to relegate perception to input, i.e. sensation, to which 'conceptualizations that are only incidentally related' are applied.[2] Such a distinction could only be valid if the percepts to which concepts are presumed to be attached are no more than registrations without conceptual weight, or if the conceptual structure of consciousness is a faculty distinct from its objects. The distinction of consciousness and objects makes an extrinsic entity of an ingredient, divides the object from its conceptual base and then, through interaction, attempts to reunite them by external contacts. This approach distances aesthetics from the psychology of everyday perception, and offers no bridge from aesthetic entities to magical, psychotic or devotional objects, not to mention the beauty of sunsets, chess moves and the sight of one's lover.

The subjectivity in a work of art was discussed by Plotinus who asked, if 'the potentiality is the substratum while the thing in actualization – the statue for example – is a combination, how are we to describe the form that has entered the bronze?'[3] He went on the say that if the relation of the potential to the actual is that of two different substrates, then 'the potential does not really become the actual: all that happens is that an actual entity takes the place of a potential.' He hinted how this might occur: 'anything that has a potentiality is actually something else, and this potentiality of the future mode of being is an existing mode.' In other words, the concept of potential is a relation to what is actual in the future of the potential, but potential can also be described as what is actual in the present. Potential unspecified is subjectively actual, another mode of existence prior to that of the final or material object.

Actualities are not connected in causal chains but emerge from potential as replacements. Otherwise, an antecedent cause that was a prior actuality would be disenfranchised from the perceptual content. In process theory, potential specifies a concept that further specifies an object.[4] The concept is the immediate past of an object, an object is the future toward which the concept is heading. The object perishes for a new concept, the concept perishes as it objectifies. The momentary past of the object, its immediate present, and the aim to actualization, are all part of the same object. Actuality is a limit, potential a direction. The inclusion of subjective phases in the objective content implies that an artwork does not consist of two portions, a conceptual part for the creative process, and an objective part that is its physical realization. The work of art is fully physical or fully mental – assuming a physicalist account of meaning and/or a mentalist account of brain process – but not half one, half the other; it is one *and/or* the other all the way through. The artistic object is that portion of the subjective that exists in the world; an actuality of creative toil that includes its creative segment which is then reincarnated in the mind of the observer. Art energizes the psychic undersurface of the objectively real in the artist, and in the aesthete.

Process and pathology

It is natural to think that objects are offered up to consciousness with the
self the spectator of a passing show, while in truth, consciousness and the
self are deposited in the course of the perception. The self is a residue of
constructs within a perception left behind as the object moves outward.
The self is laid down by phases in the object that were bypassed in the
surge to objectivity, phases that are uncovered in states of reverie, meditation and, more reliably, in cases of brain pathology. These phases, usually
traversed automatically, are recaptured in the letting-go of deep meditation, and the coming-to-the-fore of the symptoms of brain damage.

Clinical study shows that objects have a 'structure' the greater part of
which is private, concealed and differentially vulnerable to pathology. The
symptom exposes a phase in the specification of microstructure teased
apart by the injury. The process mediated by the damaged segment
'discharges' in a symptom, which is a sign of that phase in the specification
process. Within the conceptual phase of the object formation, the analysis
of perceptual deficits demonstrates a continuum from abstract category to
concrete entity to final object that unfolds over an evolutionary hierarchy
of neural systems. Symptoms arising at successive points in this hierarchy,
as inferred from pathology, reflect moments in the genesis of a perception.
In a word, brain injury displays early process by disrupting it.

In perceptual disorders, a derailment of object meaning spares object
form, and the reverse. An individual may perceive an object but fail to
recognize it, sort objects in categories without being able to identify them,
or misperceive visual objects but recognize them by touch or audition.
The finding that meanings can be disrupted yet remain submerged in
object representations gives credence to the notion that conceptual
feeling is an antecedent phase.[5] The conceptual phases leave their traces
in an anomalous content that is still part of the perception even if the final
objectivity is all that is perceived. Specifically, form perception may be
intact with a derailment of object meaning, while meaning may be
preserved with a disruption of form.

In the brain-damaged, there is usually a preference for symmetry and
simplicity of design. Severe aphasics cannot be questioned as to their
aesthetic responses, but they seem to retain an enjoyment of fine art and
music. The aesthetic response, or its manifestations in habit and taste,
survive a loss of language. In artists, there can be preservation of drawing,
often with an alteration of conceptual content. The Bulgarian artist, Zlatio
Boiyadiev, suffered a severe stroke with total aphasia, and a shift from an
accomplished social realism to a style that was more impressionistic, with
bold colors, inverted perspectives and heavy brush strokes.[6] It is my observation that a disruption of phonology tends to give alterations in perspective though drawing is otherwise preserved, while in lexical-semantic
disorders there is increased fantasy, often of a dream-like nature.

One does not need pathology to demonstrate the influence on object perception of a nonconscious conceptuality. Every object is shaped by experience. Indeed, every past experience in life is implicit in the occurrent state.[7] The more recent events are configured by the more ancient ones, with the present state most exactly configured by the one immediately prior to it. We see the effects of past experience in the familiarity of an experience repeated over time, for example, even to the unsophisticated listener, a piece of music becomes more familiar, i.e. is perceived in greater detail, with repeated listening. This effect is unconscious. In experiments, faces that are shown repeatedly to subjects are judged as more attractive even though the subjects do not recall seeing them before. Here, judgments of beauty are linked to unconscious familiarity. Indeed, familiarity is a learning-by-acquaintance of some complexity involving a nonconscious process that can alter perception even in the absence of a recognition that the object has been previously encountered. Once recognition sets in, familiarity follows as a matter of course, though this does not exhaust its explanation.[8]

Temporal objects

The clinical material is important because a specific conceptual preprocess cannot be inferred from an outcome. McGinn wrote, one cannot 'recover from a functional ascription to an organism the specific means of discharging that function in the organism in question . . . the function of an evolved characteristic does not determine the specific intrinsic nature of the characteristic'.[9] I interpret this to mean, trivially, that 'output' does not predict interior process. This might imply that there are many roads to Rome or that, from the destination, the one and only road is indetectible, or that the way to the output is part of the output, not a causal antecedent, and undiscoverable from the output alone. One can agree, at least, that the surface expression or content of a mental state reveals neither the processes supporting the expression nor those antecedent to it.

We take mental entities at 'face value', decide on their relevance, utility, standing, even though we know little or nothing of the mental processes through which they develop. In the conventional approach to aesthetic and ordinary perception, the distinction that is made between content and becoming splits off the conceptual from the objective and, in so doing, eliminates mind from the object. Certainly, mental contents, no less than made objects, can have the status of scientific entities, the truth of which can be ascertained independent of their source. In everyday life, we do not need a recipe to enjoy a meal, and we can usually distinguish rational thoughts from fantasies without submitting the content to depth analysis. But, we cannot derive the *laws* of cooking from what we have eaten, nor of mentation from its products, nor even from a list of ingredients or components. This would be like explaining how a television works by looking at the wires and transistors or the picture on the screen. The laws of thought cannot be extracted from the contents of thought for the laws describe how the contents come into existence.

Such processes are elusive. Psychologists assume they refer to an aggregation or recombination of elemental part-functions, which are obtained by 'deconstructing' the object into its presumed constituents. These physiological operations, and the little agencies through which they act, such as feature or motion detectors, color-coded cells, etc., are conceived as neural devices that subserve a variety of 'dumb' routines, all of which, through a process of assemblage, go into the construction of an object, though it is unclear in what sense they may be said to eventuate in an object or be ingredient in it. The part-functions are assumed to have a causal role in the production of a perception, but we are unaware of the assemblage, how is it accomplished, how memory and thought enter the perception, how the process 'jumps' from neural routine to mental state and whether the effect of each element is concluded when the perception is achieved or is instantaneous, i.e. whether perceptions are outcomes or sums, and so on.

It must be emphasized that in cases of brain damage, it is unusual to encounter a defective part-function of this type. A loss of line or angle detection, a motionless object, the absence of a single color, do not appear to occur with focal lesions, other than as artifacts of the expectations of the investigator. What we do see, for example, are a variety of qualitative alterations that fluctuate in perception. These fluctuations are linked to aspects of the personality. The condition is classified according to the major impact of pathology, not the loss of specific components. When a focal defect does appear, it affects the perception momentarily, uniformly, not in the patchwork and sustained manner anticipated by a dissection of particulate elements.

Studies of recovery help to clarify this problem. In cortical blindness, the return of vision is systematic and predictable. First, a diffuse brightness sensation returns, then gradients of brightness, then a vague sense of motion, or apparent motion, followed by size, shape and depth. The first chromatic color to return is a diffuse, filmy, unsaturated redness. Color tends to precede form in an orderly sequence. The pattern of recovery is regular from one phase to another, within and across cases. This argues against the damage of specific components. In the resolution of aphasia, recovery is also holistic, involving levels of linguistic realization, not a return of piecemeal functions, as would be expected if damage produced a random destruction of a mosaic of elements.

The clinical evidence that phases in object realization are revealed in the form of symptoms reinforces the conclusion that objects are not stationary entities or slices in the stream of time but have temporal extension. Put differently, phases in the actualization, from initial to final, are not preparatory to the object but *are* the object. It is even unclear whether a *perceptible entity* is realizable in the fraction of a second that is a single act of perception. An object requires both the phase-sequence within a mental state, and a duration over mental states, to be perceived as self-identical. Elsewhere, I have referred to the within-state transition as implicit or authentic change, and the state-to-state transition as explicit or apparent change. The first, the unfolding of a single mental state, creates an epochal object, the other, the 'glue of passage', sums the object across its several replacements.

Once an object is understood as the 'sum' of a temporal series, within and across cognitions, the problem arises as to how the sequence is to be demarcated. The extrinsic sequence is ostensibly non-problematic. The brain, so the argument goes, simply fuses successive objects like a picture strip. There are difficulties with this interpretation but for the purpose of this discussion they can be ignored. The intrinsic series, however, presents a greater problem. The phases constituting an object can, in principle, be stretched to include the inception (arising) of an object in the mind of the observer and its termination (perishing) in external space, when it is replaced by another perception. If the inception includes conceptual phases, where is a boundary to be drawn between an object and a concept?

The only response to this question is that the temporal extension of an object must incorporate all of the (neural, psychic) phases in its satisfaction. The physical referents of the object are the physical worlds implicit in its actualization. These worlds constrain the object as it becomes real. At every phase, the world, in the form of brain process, combines external and internal constraints to guide the object along. An external constraint, e.g. sensation, is no more or less defining of the resultant configuration than the internal process, e.g. habit, on which it acts. An object is the set of contrasts that survives this process. The configuration as a whole – the contrast with its adjacent surround – and the parts that individuate, are apprehended jointly. Whitehead wrote, 'esthetic experience is feeling arising out of the realization of contrast under identity'.[10] A contrast isolates a particular in a greater inclusiveness. A constraint is a type of contrast that frames a possibility. This boundary is a compromise of potential with its limits, of what is possible with what is permissible, of what can be thought of with what is real.

Music

Our theory of the world is conditioned on the perception of visual space. Indeed, it is difficult to even conceive of a thing that cannot be visualized.[11] Visual phenomena are fundamentally temporal, grounded in becoming, but experienced in a spatial mode. The spatial mode is the mode of existence attributed to a non-cognitive entity and then transferred to the entity when it is being observed. A painting is a cognition that leaves the mind of the artist and enters the mind of the observer. When it is not present in a perception, it is a pattern of energy, like quartz or water. Its temporality enlarges in an occasion of its enjoyment. That is, subjectivity expands the temporal relations concealed in the physical object or artwork. What, then, of art-forms that come into existence *only* when in performance, where a translation from a spatial to a temporal mode is the dominant condition of experience, such as music?

To the casual observer, objects and tones, notes and music, could not be more different, as different as space and time, as the strings of a violin from the sounds they convey. We are constantly reminded of the contrast between the evanescence of sound and the persistence and slow decay of objects. An object is perceived as an entity continuous across occasions. Changes of the object – growth, decay, interaction – are perceived not in terms of a changed object but as changes the object undergoes. The object, it seems, would remain unchanged if change were to cease, as if change were an event in the career of the object. Time is an agent of destruction that erodes the substance of what would otherwise be a continuous existence. An object will be dust in a million years. The attrition seems to be the effect of time. Were it not for this effect, we think, the object might last forever.

In contrast, change as a transition across tones is the essence of music. Tonal identity is sacrificed for the sake of melodic continuity. In visual perception, change is filtered out for the sake of object stability, whereas in music, the melodic phrase or the structure of the work as a whole spreads out serially in the surface tones and is recovered by the listener according to his musical experience. There is no stable object, the *tone*, the *music*, that undergoes change. Music seems to run *in* time, the notes seize time and exploit it, while objects are at the mercy of time, time seizing them and wearing them down. Music, Kant said, is the 'art of time', in which temporal order is employed as a formal property.[12] What is temporal order other than an articulation by events of some duration? The order does not create the duration, rather, the process generating the events creates the order, which then objectifies in time-awareness. Each percept increments an individual consciousness.[13]

When we listen to music we do not hear a sequence of acoustic waves – the sound impulses that strike the ear – but a phenomenal entity, an auditory image, that develops in the mental state. Tones are more like images than perceptions. One could say, we hear the music in our souls before we hear it in the world. The perceived tones arise out of the images of prior tones, sound pictures that 'take time' to develop. For music, an echo of the tone-image must recur after its cessation. The tone-image retains its serial position in relation to occurrent tones, inheriting and replacing the preceding image, and stacked in the mind like tea gardens. The succession of phases separates the tones to prevent a muddle of images. The difference between successive tones is cancelled in their replacement and retained as an apprehension of sequence in a musical phrase. The difference that is *cancelled* is the prior change erased in the becoming of the present. The difference that is *retained* is the comparison of a prior image with a present actuality as a limit to which antecedent states are revived or decay.

In process theory, an object is not a linear solid with change at its forward edge but the relic of change left behind in its cyclical recurrence. It *reappears* each moment in more or less the same form. We are convinced by the similarity of perceptual frames that the continuity of the object is independent of the observer when, in fact, its replication in the mind is the basis for its stability. An object is changing like music, but our awareness of the change is obscured by the similarity of replicates. For example, the radio on the table appears as a solid object that persists through time, quite different from the music that emanates from it, which is a dynamic pattern of fluctuant sound. This difference reflects the recurrence of similar images that fuse to form an object with a continuous existence, in contrast to music, which is the recurrence of disparate images over a finite set of iterations. The critical difference between vision and audition has to do with the discrepancy of objects in adjacent moments of perception. The replicates of objects are similar, those of music variable. Music that does not vary, or is densely repetitive, a sequence of self-similar replicates, is like a continuous object.

The commonality of objects and sounds is seen in the resemblance of music to a visual hallucination or dream in fluid transformation. If an object were to undergo a rapid change, say a cup that contracts, flattens or expands to become a glass, a saucer or a bowl, the change from one shape to another would be so unexpected we would not be able to say what object it is we are looking at, or even if what we are looking at is an object, since objects do not behave in this way. The definition of an object would begin to loosen. If an observer was accustomed to seeing objects in continuous transformation, the perception of such objects might be construed as a form of music. An object is like a series of tones over time, an accumulation that can be viewed as a type of spatial music, a melody in which the tones are more or less identical. Goethe may have had something like this in mind when he described a complex object – architecture – as frozen music.

Conversely, music is a liquid object, the disparity between replicates giving the perception of a sequence in continuous change. Still, structure is perceived at the level of the phrase, the movement or the work as a whole. For a musical phrase to be enjoyed, the sequence must be heard within a whole unit or duration of time. The tone sequence is carved out of the present allowing the tones to be heard both in sequence and all at once, what amounts to the incrementation of a virtual simultaneity, or a derivation of tonal parts out of melodic wholes. The specious present encloses a succession of tones derived from antecedent wholes, much as a visual image is derived from a succession of perspectives. Tones are objects in auditory space isolated by figural contrast. The context around the tone – 'horizontal' in melodic relations, 'vertical' in harmony – depends on intervals in the mind's auditory space. The melodic structure thickens with the frequency of harmonic shifts as the compounding of polyphonic planes lays down strands in the melody.

In music, the structure of the work is primary, the impression of motion an illusion. There is no movement, no change in position. The pitch of a tone is a point in auditory space.[14] An interval between points is an illusory 'distance' since a point in the immediate past exists in memory when the ensuing tone is heard. The transition from one point to another in subjective time is the melodic movement. The feeling of motion is created by tones going nowhere. The tone replaces itself.

The order in which the notes of a score are played is a logical progression. The notes are simultaneous in the score, yet played and heard sequentially. In the translation from the score to the mind of the listener, they traverse intermediate holistic phases. The duration of the present can be viewed as a gestalt-like whole that encloses the tones in succession. In fact, the seriality articulates each momentary present as it is renewed in overlapping waves. The serialization of the tones in duration retains a spatial character as a projected two-dimensional line. The notion of a musical line, or the movement of the piece from beginning to end along a line in time, Bergson argued, is a spatial image of a purely temporal process. I would say the segmentation of duration into succession creates time awareness. When we give ourselves up to the music, the succession becomes a subjective time of music. Hegel wrote of the imposition of musical time on the time of subjectivity.[15]

Commonality of sound and object

Music is perceived 'in the head', objects are in the world. At a concert, when I close my eyes the sound is referred to the proscenium, but gradually it moves to a locus in intrapersonal space. The sound is now in my head. Through an effort of the (visual) imagination, it can again be projected outward. The mind extends a visual space beyond the body to provide an extrapersonal environment in which the tones can be located. The localization of sounds is parasitic on visual space. Sound has no space of its own. Disorders of visual space with intact auditory brain regions have a disruption of sound localization. There is no external space for the sounds to localize in. The space of the congenitally blind is a kinaesthetic field that extends to the perimeter of limb action. We see the dependence of audition on vision in the mapping of speech to oral movements, or the synchronization of musical sound with instruments. The concordance of objects and sounds requires not only an object for the sound to exteriorize onto, but a common organization and timing of the modalities for the synchrony within a perceptual moment.

The fundamental unity of the modalities is illustrated by the ability of gifted musicians to hear a score on reading it, or to write the music down after hearing a single performance. For Brahms, an armchair and a score at home were preferable to a seat at the opera. I have been told of linguists so proficient with sound spectrograms they could hear speech on reading the wave patterns. In the phenomenon of *écho de la lecture*, psychotics hear words they are reading as an echo a moment after the word has been read.[16] Inner speech is excited to a kind of involuntary imagery. Verbal perception is activated by reading words. Is this comparable to musical perception being activated by reading notes when a gifted musician hears a score in his head?

Cases of synaesthesia for music and colors – a color specific to a key or an instrument, for example seeing the color blue on hearing a violin, or a specificity of colors to speech sounds, for example seeing the color yellow on hearing the sound *ba* – also signify a close bond between the auditory and the visual modes.[17] These effects are most prominent for purely visual features (letters, musical notation, visual wave forms, colors) that, unlike ordinary objects, are specific to the visual modality and are not reinforced by somaesthetic experience. Synaesthesia is usually interpreted in terms of an association or transfer from one modality to another. I think that the specificity of colors and sounds derives from their common origin in a unitary matrix that forecasts the individual modalities. Synaesthesia is a reminder of the unimodal foundations of the varieties of perceptual experience.[18]

Visual signing in the deaf is mediated by auditory (language) cortex. With congenital damage to auditory or visual cortex, functional preservation of hearing and sight may occur through other cortical zones. The bond between vision and audition is confirmed by the difficulty that occurs in lip-reading with damage to auditory cortex, or the confusion that results when speech and oral movements are thrown into dyssynchrony by delayed auditory feedback. It is, after all, the same world out there for the ears and the eyes. Music is part of the cognized world. Music is clearly in the world like any other object, especially when I am not really listening. When I attend to objects, I do not hear the music. When I really listen to the music, the world disappears. But the world is ultimately one. The distinctions of sight and sound, of what is written and what is heard, of act and percept, are the surface tokens – what survives in consciousness – of conceptual feeling distributing into a given cognitive domain, into auditory perception for music, into the visual or tactile (plastic) arts for painting and sculpture, into action for dance, language for conversation and literature, and so on.

On this view, a poem, a painting and a sonata realize a common deep form.[19] The conceptual is most prominent in the poem, feeling in the

sonata. Linguistic categories are learned, object categories are naturally
acquired, musical concepts develop as a reward to careful study. The
emotive in music dominates the conceptual; with words and objects it is
the reverse. The ineffability of music, the 'thoughts too deep for tears', and
the non-intentional moods it evokes, trace to the affective background of
pre-lexical concepts. While there is some uniformity of emotional
response to a work across occasions and listeners, the music, the emotion,
the listener, are never exactly the same. The mind is differently prepared.
Keats was partly right. The sweetness of melodies heard is enhanced by
concepts unheard to make the music sweeter.[20]

Content and meaning

Without subjectivity, a poem is a scrawl of marks, a painting, a blotch of
color. The *Eroica* played on the moon is noise. These become artworks,
again, when they pass through a cognition attuned to their significance.
The difference is the presence of meaning. A work of art may seem to have
a meaning for the subject to decipher or it may suggest a meaning for the
subject to provide. Meaning is knowledge in the service of value without
which a perception is a neutral datum. The valuation in meaning is the
subjective in knowledge. It transforms concepts to personal beliefs, to the
point where the subjectivity of knowing, the *belongingness* of what is
known, has a greater immediacy for the subject than the content the
meaning is about. Like knowledge, meaning can be implicit or conscious,
it can be assigned to the subject, as a content in the mind, or located in the
object, as a content in the world, that is, to an interpretation applied by
the subject or an external commodity gathered up by the subject and put
to work. An object dissociated from its meaning is displaced to the
external world, or a locus intermediate between reality on the outer side
and mentality on the inner side as a screen between conceptual feeling
and the noumenal universe.

When we are young, we seek meaning in the world where knowledge
seems to be located. As we mature, and our 'stock of knowledge' expands,
i.e. as mind is enriched by experience, we come to realize that meanings
after all are creations of the subject, to be generated and sought after in his
own nature. Later, we may realize that we only discover the meanings that
nature provides. That is, meaning does not move inward from the world
to the mind, nor outward as a mental factor attached to objects that otherwise would be meaningless, rather it is ingredient in the conceptual
relations that specify objects and the mental representations through
which they are enjoyed. Meanings inhere in the configural properties of
cognitions and deposit as the subjectivity of the onlooker prior to the
realization of objective or linguistic form.

Accordingly, an object that seems meaningless, such as a foreign
language or a nonsense shape, still traverses a phase of meaning-relations,

if only by the recognition of a lack of meaning, i.e. the meaning of a meaningless object is impoverished, not lacking, the configural biases guiding the process leaving no more than a rudimentary knowledge base, either because the object is strange, or of insufficient interest to evoke a meaning in the observer, or because the conceptual system of the observer is insufficiently developed to respond to an otherwise meaningful object. In becoming actual, every object travels the same path. There is a spectrum from contents that seem meaningless, because of inadequate knowledge, to contents in which meanings are felt to be interpretations, to contents that seem to provide (constrain) the meanings which the subject attaches to them. In all of these instances, however, we are dealing with gradations in conceptual feeling evoked as an object materializes under the constraints of form. The conceptual base of the object, i.e. whatever the subject brings to the perception, leaves its traces, subtly, in the tacit recognition, identification, classification or value that guide the object to its destination and partition it from the rest of nature. Conversely, the elicitation of knowledge or reminiscence by an object, or a shift of attention from the object to its contextual background, signals a retreat to the abstract categories that were sacrificed when a concrete object was their goal.

Art and nature

The view that meaning inheres in the configural properties of developing objects entails that even without a subject an object is not meaningless. An object hewn from a subject is still the objectified portion of its own subjective phase.[21] In all this, we are speaking of external objects such as trees and chairs that are assumed to exist independent of our percepts. Such entities, in themselves non-cognitive, could be described as meaningless only if, like the objects of scientific thought, they are conceived as physical entities supplemented by mind-dependent properties. If meaning is generated by the intrinsic spatio-temporal relatedness of non-cognitive entities, such as the relation of spatial wholes to parts, or durations to temporal increments, and if a linguistic operation is not obligatory to meaning and, finally, if the intentional relation to otherness is equally an aim toward self or object-realization, i.e. if the intentional is a relation *within* the object, interpreted as a single completed state or entity rather than a vector from mind to external objects, the seeds of meaning and intentionality in non-cognitive entities would qualify them as primitive conceptual systems.

If so, meaning, or conceptual feeling, would arise, primitively, in non-cognitive entities as the subjective phase of objects undergoing temporal completion, or in human cognition, in the withdrawal from the world of actuality to its derivational phases. One can ask, poetically, with Menander, whether nature or art is the plagiarist, but ordinarily we are able to distinguish natural objects from artworks. The central features of artistic creation – purpose, freedom, subjectivity, aboutness – are not found in nature, or so it is argued. Nature is functional, its design, unlike art, is adapted to survival. Subtract the constraints of evolution and nature is chaos. Subtract the intentional and an artwork is a random assortment. Nature is what it is, mechanically, causally, inevitably, it does not mean or point to something. For most of us, the works of nature are not comparable to works of art, unless God is an artist and a tree is conceived as an intentional product of God's own agency. Yet the true artist, in his art, summons up the power and generativity of nature; he reaches into his own *nature*, to the patterns of mental process that are continuous with those of nature, and transforms it into art.

We see this transition in the magical objects of the primitive, which for some are not even artworks for they violate many of the canons of aesthetic perception. It is ironic, therefore, that they can be appreciated, *aesthetically*, only by those with a theory of art outside the culture that produced them. For the primitives, however, their artworks, i.e. *artifacts*, do not so much express intentions as assign intentional properties to nature through a 'projection' of the agency of the *artisan* into the realized object. Primitive concepts leave the mind of the artist and take up residence in the artwork. The anonymity and sacramental quality of primitive art derive from this projection. The artist is less an individual creator than the voice of the clan or community, the work, less a signature of the artist than a zone of mentality within subjective nature.[22] From the primitive, we learn that artworks fall somewhere in a process of becoming independent of a definition of what an artwork is supposed to be, and that the continuum from nature to art, like that from ordinary objects to aesthetic perception, is a deep fact about nature and the human mind.

In fact, the flow in art from mind to nature, and back again in the process of creation and enjoyment, is observed over the historical period in the progression from its origins as an image of nature, in primitive art, where agency and feeling are centered in the artwork, to imitation, where the artwork, though divested of agency and feeling, is still close to the external in its effort to represent an image of the real, to the exploration of the mind of the artist as a container of its own objects. An object that fully actualizes the conceptual content behind it is a psychic entity, less an artwork than a *talisman*. The autonomy of an *ordinary object* is achieved by usurping or transforming the ancestral concepts out of which it took shape. An *artwork* is an object that leaves the conceptual behind as a wake of feeling activated in its transit to the world. This shift in subjectivity from a fixation in nature, to a subjectivity that is a spectator of concrete entities in the world, to a subjectivity that is absorbed in the creative process behind the artwork is, really, a relative accentuation of one or another phase in a continuous 'sheet of mentation'.

The objectification of concepts

Objects are implicit Rorschachs. The lure to meaning so characteristic of human cognition announces the conceptual feeling that is driving what on first approximation seem to be naive perceptions. Such objects are perceived as barren of intrinsic meaning because they remain more or less constant in form while their conceptual foundations undergo an expansion brought on by repeated contacts. Similarly, an artwork does not appear to undergo a transformation even with an increase in knowledge of the work, though to be sure, we may see more in the object with each new exposure. When the artwork does not provide clues to interpretation, some history and explication are needed to elicit the context from which the 'naive' representations emerged.

The knowledge that goes into an aesthetic experience may be felt as secondary for the same reason as in ordinary perception, that abstract concepts seem to develop out of empirical ones. The interpretation of the concrete or empirical as prior – thus, inferior – to the abstract is responsible for the historical error that abstract concepts are *constructions of thought* or *pure conceptions* that are brought to bear on low-level object perceptions. This leads to the contrast between basic and aesthetic objects, or between the latter and intellectual objects, or the opposition of concepts and objects, or reason and logic with unmediated intuitions. The history of aesthetics is replete with descriptions of cognitive abilities that are scaled from the ethereal to the bestial, in which the act of creation is conceived as an enterprise somewhere between mysticism or spirituality at one extreme, and childish play at the other, the mind assumed to be a kind of edifice with lower and higher floors in which, depending on the writer, art inhabits either the cellar or the penthouse.[23]

This is not the conclusion of clinical studies, which find abstract and concrete concepts to be the expressions of categories of graded, qualitative uniqueness that lead from those of the widest scope, and indefiniteness, to those of the narrowest application. This is also true for feelings, which resolve from the pressures of the basic drives to the desires and the partial affects or affect-ideas. Preference is the child of desire, desire mitigates drive and 'intentionalizes' in a particular object.[24] An aesthetic object is an image *en route* from an abstract concept to a concrete perception, accompanied by desires or feeling-tones that also are intermediate between the drives and exteriorized affect, i.e. the feeling of reality and value in the object. In both evolution and mental process, the individuation of the particular from the general, the concrete from the abstract – objects from concepts, aesthetic concepts (images, symbols) from abstract ones – is a partition of constructs of greater inclusiveness and potentiality to those successively more delimited, final and committed.

Thus, growth in knowledge can be viewed, not as a graft of reason to the repertoire of perception but as a propagation, appropriation or cognitive 'spandrel' of the conceptual underpinnings of ordinary objects where it was incipient, unnoticed and more or less automatic in the initial encounter. The attributes of form, proportion, timing, rhythm, etc., and the categories of object relations and personal memory that come into play in the perception of common objects, undergo an elaboration *from within* to categories that ground the evaluation of aesthetic objects and their psychological interpretations. These aesthetic valuations, including affective tonality and the relatedness or groundedness to objects, derive from, or can be further distilled to, philosophical concepts which are at a still greater remove from concrete reality. The aesthetic image is midway between the detachment of reflection and the immediacy of perception. For this reason, it vivifies the concealed undersurface of objects while still remaining bound to an object experience. Conceptual feeling is less apparent in ordinary objects than in artworks, yet is no less decisive in shaping our sense of what an object is, whether a thumbtack or a Picasso, whether the sounds of traffic or the *Eroica*.

In the relation of artist to artwork, or that of abstract concepts to empirical objects, aesthetic objects more than everyday artifacts show a heightened sense of interioricity or subjective individuality.[25] The penetration of objects by a felt conceptuality, the sense that there is more to the artwork than its overt material, and the harmony and coherence in (the perception of) the work – a sign that aesthetic entities are adapted to the domain of concepts, not objects – point to possibilities undeclared in the final content. This is Kant's 'aesthetic universality' of art, in contrast to the 'objective reality' of logic and science. For Cassirer, the potentiality of aesthetic experience was 'pregnant with infinite possibilities which remain unrealized in ordinary sense experience.'[26] This manifold of potential has not, as in ordinary perception, been given up in the realization of a concrete form but persists, immediate in the object, unexhausted in the detail of the representation, as a layer of thought behind the surface appearance.

The unity of conceptual feeling in a work of art is a sign of its depth. In minds of extraordinary richness, concepts arise, anterior to specification, prior to consciousness, at a presuppositional core, fundamental to personality, that provide an organic unity to the content that individuates in the artwork. In lesser artists, the conceptual origin does not penetrate to this depth. The art is not *authentic*, and the separate part-concepts serve as a makeshift source of a work that seems fragmentary, the elements lacking the unity that would have been provided by access to more generic constructs.

Objectivity and function

A painting can always be given a function, a lively pattern on a wall, a tranquillizer for people who are disturbed, a tray, a dart board, moral teaching, 'food for the soul', but the function is inessential to its aesthetic.

A functional entity, such as the gleaming chrome engine of a luxury car is, one might say, a 'work of art', but its function gets in the way of its aesthetic power. Unlike the objects of nature, an artifact exhibits all the attributes of human agency but its function situates it in the world of objects rather than artworks. As Collingwood wrote, it becomes an object for consumption, not contemplation.[27] In aesthetic objects, the use or purpose is suspended for the sake of the conceptual role. The mentality in the object is retained by an incomplete objectification. The inutility is the lack of adaptation to the world of function. This lack of adaptation or functional specificity is experienced as an indefiniteness. Since the function or meaning of the object is incompletely specified, it retains the potential and ambiguity of early cognition.[28] The perception, in completing its development, has the capacity to arouse the not-yet-committed of the interior space of privacy, anticipation, choice, possibility, that precede the irrevocability of the actual.

A suspension of functional role unaccompanied by an alteration of intentionality – the for-itself of art, the detachment or inutility, nonfunctionality and aesthetic indifference of Kant – are accomplished by transforming a functional to a conceptual intent, i.e. by shifting intention to an intrapsychic aim at the expense of external targets. A porcelain vase is an artwork that has a function, but the function is secondary. The conceptual intent is primary. The fact that the object *retains* a function is less important than that the function is no longer the goal of the intentionality behind it. In the decorative arts, a reminder of function can accompany a response to the beauty of the object, but at the cost of some aesthetic power. The object is beautiful, but do we respond to it as to a work of art? One may enjoy or wish to possess a beautiful object, but the joy or wish does not seem to be part of an aesthetic feeling. Beauty entices, beckons, seduces – the prophet of truth, von Hartmann wrote – but in aesthetics, as in life, it arouses the very desires it is unable to satisfy.

The judgment of beauty is a response to the coherence of outer form, aesthetic judgment is based on a coherence of conceptual feeling. The former is entrenched in the objective, it is felt by all and does not obligate interpretation, the latter is interpretative and develops with experience and learning. The beauty of a sunset is not enhanced by a grounding in the study of optics. A growing intimacy with one who is loved does not, necessarily, augment the beauty of the beloved in the eyes of her lover nor intensify the passion of a first kiss. The continuity of the conceptual and objective explains the association of aesthetics with beauty by the inseparability of the inner and outer aspects of perception. The fit of concepts or their features and the derivation of featural content out of conceptual feeling accounts for the interweaving of aesthetic valuations and judgments of beauty.

So it is that a work of art seeks objectivity but ought not expect to achieve it. The fully objective is toxic to the creative. A concept discharges its potential, then perishes once objectivity is achieved. Art awakens and revives those formative concepts still undepleted by what was given to actuality. Consider a painting so realistic that one expects, with the artist, that a bee will alight on a flower. Or, a piece of music that attempts to duplicate the sound of a train, or the song of a bird. In such works, the artist avoids the accusation of mere imitation, or a fatuous mastery of technique, by the use of illusion to illustrate the real.[29] Realist art is more than photographic and this difference is the basis of its aesthetic interest.[30] Similarly, poetry may violate grammatical rules or semantic conventions to elicit meanings in novel arrangements. Meaning veers to anomaly, stretching the semantic fields of words, or creating new ones. Metaphor is not logical. A true fact is a boring poem.[31] The real conceals the poetic which is the possibility of something unexpected. Coleridge wrote of the 'suspension of disbelief' that makes art possible. The suspension is the prominence of the prelogical. Art submits to logic at the cost of novelty and surprise. Art, it has been said, is the relation of appearance to reality. An appearance is unreal only in comparison to a greater – more concrete or objective – realness. Perhaps one could say there is more of appearance in art, and more of reality in perception, but it is only a matter of emphasis.

The uniqueness or rarity of the object is important, though trivially, in heightening aesthetic interest and reducing the impact of function. Nature replicates in eternal cycles, art originates in fleeting moments. A function implies a pre-existing plan, art suggests the freely creative. We may overlook utility if an object is sufficiently rare. If all but one chair in the world disappeared, that one would be a treasured artwork. The relative strength of pragmatic to conceptual intentions, and the felt reciprocality of the creative and the useful, are more decisive when there are many instances of the same object. For better or worse, these effects can be exploited in different art forms. In dance, in rock, in middle eastern, Indian and some modern music, e.g. the works of Phillip Glass, repetition can induce states of restlessness, elation, transport and dreamy intoxication. In the visual arts, an excess of exemplars in a single format, say, the calculated iterations of Andy Warhol or the deeper, limpid rectangles of Mark Rothko, tend to saturate the viewer with a sense of automation, ease and externality that can either subdue or heighten an interior response. Auditory and visual perception may have different effects. Vision cannot go on without incessant change, audition demands stability. There is nothing so tiresome as to watch the whirling dervishes, yet no music is more conducive to trance.

Artist and observer

Perception is an active process with an object as the outcome. To say the observer must recreate the artwork is to agree that the mind is not a passive receptacle but that objects are actively generated. Dewey wrote, 'to perceive, a beholder must create his own experience. And his creation must include relations comparable to those which the original producer underwent . . . Without an act of recreation the object is not perceived as a work of art.'[32] This is unobjectionable as far as it goes, but it does not go very far. What exactly is in the mind of the artist when the work is produced? Bosanquet wrote, 'if we could get at the explicit being of a creative or accumulative mind at any limited moment, we might find little or nothing there', likening it to a pencil point that leaves behind 'a splendid and intricate pattern, but has never had in it at any instant any appreciable portion of the design.'[33]

The requirement that an aesthetic perception correspond with the state of the artist's mind before vetted as genuine demands, for example, a mental state comparable to that of Beethoven to appreciate his music, which excludes all humanity from such an experience and assumes that, even were one to generate such a state, the only thing preventing composition in the manner of Beethoven is lack of compositional *skill*, as if conceptual feeling were independent of its expression in an act or object. Conceptual feeling is accentuated in the attempt to appropriate the creative energies of the artist in an act of imaginative fusion.[34] The artwork does not reproduce in others the artist's state of mind but arouses concepts and feelings of some generality that touch on a common humanity.

The sublime

Ordinary objects limit aesthetic perception at one extreme, the experience of the sublime at the other. The feeling of sublimity may owe to a simultaneous apprehension of these limits, the immediacy of an overwhelming object at the outer rim of perception, the expansiveness of potential at its base. The sensibilities that converge in this experience include the relations in space and time, an identification of becoming in mind with that in nature, individuality in duration, and a oneness with the natural order in all its terror and tranquillity.[35]

A vastness impenetrable to reason yet comprehensible in the imagination has, since Kant, been claimed to be essential to the feeling of the sublime.[36] The artist also captures the interior motion of momentary consciousness in relation to the power of creative advance. The immensity of nature in relation to the insignificance of the individual is the spatial analogue of the temporal embedding of the evanescent in the eternal, the view from Olympus with the fragile human perspective. Aesthetic experience is a capsule of the sublime, where the artwork is perceived as a fragment of an imaginative whole, potential with its realized elements, tones in phrases, temporal parts in durations. The part/whole relation in time and space, central to aesthetic feeling, expands outward to embrace the limits of time and nature, and inward to the local and the fleeting. The permutations of entities developing in the mind, changeless in perception and perishing once perceived, tinges the experience with a melancholy awareness of the finality of passage and the fading of brief existents.

The sublime, then, is an experience of part and whole in time and in space. In its temporal manifestation, it is the compresence of the personal now with an eternal present, or the vulnerability of linear time embedded in the deep cyclicity of change. In its spatial manifestation, it is the immersion of the individual self in the immensity of nature or an individual *oneness* in an impersonal *many*. That is, it is the contemplation of an individual *time* and *place* in the spatiotemporal *whole* that surrounds it, from which it arises, to which it returns. We die as leaves on the tree of matter. All creatures, inanimate and living, a bird, a bower, a ripple in the stream, the clouds above, the ragged cliffs, inhabit the same world. Wordsworth wrote, 'every flower enjoys the air it breathes'. Creation is thematic, in the very nature coursing through our veins, the flowers and the hills, the torrents and the winds. In contemplation, all distinctions melt away. The enormity, the manifold, the diversity, are then felt to issue from a single incomprehensible root sensed in an act of beholding.

Notes to Chapters

Notes to Chapter 1

[1]For the history of the idea, see Heinz Werner, 'Microgenesis and Aphasia', Journal of Abnormal Social Psychology 52 (1956), 347–353. See also: Robert Hanlon and J.W. Brown, 'Microgenesis: Historical Review and Current Studies', Brain Organization of Language and Cognitive Processes, edited by Alfredo Ardila and Feggy Ostrosky (New York: Plenum, 1989) and Cognitive Microgenesis: A Neuropsychological Perspective, edited by Robert Hanlon (New York: Springer-Verlag, 1991). On neuropsychological data in support of the theory, see my Life of the Mind (hereafter cited as LM).

[2]See the article by Paul MacLean relating microgenesis to fractal geometry, 'Neofronto-cerebellar Evolution in Regard to Computation and Prediction: Some Fractal Aspects of Microgenesis', in Hanlon, see note 1.

[3]F.B. Wallack, The Epocal Nature of Process in Whitehead's Metaphysics (cited as ENP).

[4]Jason W. Brown, Self and Process (cited as SP), Time, Will, and Mental Process (cited as TWMP).

[5]For example, Edward Pols, Whitehead's Metaphysics (Carbondale, IL: Southern Illinois University Press, 1967). See also the discussion in William Christian, 'Some Aspects of Whitehead's Metaphysics', Explorations in Whitehead's Philosophy, edited by Lewis Ford and George Kline (New York: Fordham University Press, 1983), 31–44. See the discussion in Lewis Ford, 'Can Whitehead's God be Rescued from Process Theism?' Logic, God and Metaphysics, edited by J. Harris (Netherlands: Kluwer, 1992), 19–39.

[6]See the discussion in Paul Schmidt, Perception and Cosmology in Whitehead's Philosophy (New York: Rutgers University Press, 1967).

Notes to Chapter 2

[1]Satkari Mookerjee, S. (1935) The Buddhist Philosophy of Universal Flux, University of Calcutta, Calcutta.

[2]The conjunction of successive moments through contiguity, causal efficacy, the persistence of the cause in the effect, the 'thickness' of the elements, the properties of onset and offset – existence/non-existence couplets – identity across instants, causation, and so on, are topics that have been extensively debated and will not be discussed at length in this chapter.

[3]Stcherbatsky, Th. (1923) The Central Conception of Buddhism and the Meaning of the Word 'Dharma'. Royal Asiatic Society, London.

[4]Murti, T.R.V. (1955) The Central Philosophy of Buddhism. George Allen and Unwin, London, 71.

[5]For example, in Kalupahana, D. (1975) Causality: the Central Philosophy of Buddhism. University of Hawaii Press, Honolulu. And, elsewhere, the problem with 'identifying consciousness with the passing psychical states (is that in the absence of a distinct mental entity behind the states) consciousness has been reduced to a congeries of momentary conscious units having no real nexus between'. (Mookerjee, S. (1935) The Buddhist Philosophy of Universal Flux University of Calcutta, Calcutta, 342).

[6]See the discussion in Garfield, J. (1995) The Fundamental Wisdom of the Middle Way: Nāgārjuna's Mūlamadhyamakakārikā. Oxford University Press, New York.

[7]Candrakirti writes that when a cognition is manifest from a potential then there is a 'dependence on a reciprocal object' that in virtue of the dependence does not exist. The critique of before and after, and the non-existence of dependency, has less force if the events are conceived as successive phases in becoming, since the non-temporality of becoming dissolves the problems of priority and relational dependency.

[8]The insistence that emptiness is viewless because it is a lifepath that excludes or transcends all other views, though resonant with the earliest concepts of Mādhyamika, has the unfortunate effect of restricting discourse to exegesis when there is much in Buddhist philosophy that could profitably be applied to scientific thinking. In this respect, I find myself closer in spirit to Murti than to his many critics. See Huntington, C (1989) The Emptiness of Emptiness, University of Hawaii Press, Honolulu.

[9]Estimated at 10^{-24} seconds by Whitrow, G. (1961) The Natural Philosophy of Time. Thomas Nelson and Sons, London.

[10]Ramanan, K.V. (1966) Nāgārjuna's Philosophy. Charles Tuttle, Vermont.

[11]Stcherbatsky, Th. (1993) Buddhist Logic Vol. 1. Motilal Banarsidass, Delhi, 84 et seq. reprint 1930 edition.

[12]Hartshorne, C. (1975) Whitehead's differences from Buddhism. Philosophy East and West 25: 407–426.

[13]See: Odin, S. (1982) Process Metaphysics and Hua-Yen Buddhism. SUNY Press, Albany.

[14]The term state is unfelicitous for it implies a static block or slice of time. It is used to describe the full set of kinetic phases in the dynamic 'module' of a becoming sequence. The term phase refers to a transitional segment in a given state.

[15]Murti, T. (1955) The Central Philosophy of Buddhism. George Allen and Unwin, London, 73.

[16]See: Garfield, J. (1994) Dependent arising and the emptiness of emptiness: why did Nāgārjuna start with causation? Philosophy East and West 44: 219–250.

[17]Mehta, P. (1956) Early Indian Religious Thought. Luzac and Co., London, 230.

[18]A theory of how duration arises in the mental state, and the relation of the duration of the present to the stability of mental entities and external objects, has been discussed in previous books (see Brown, Self and Process (1991) and Time, Will and Mental Process(1996)).

[19]Mehta, Early Indian Religious Thought, 347.

Notes to Chapter 4

[1]Ironically, Freud wrote to Binswanger that Wernicke was 'an interesting example of the poverty of scientific thought . . . and could not help dissecting the soul as he had the brain' (Binswanger, p. 36). Freud, of course, went on to do precisely that, though after the Project there were no further attempts at neurological correlation.

[2]Freud's ideas on association and imagery were also influenced by J.S. Mill (1880), whom he translated to German, and by the work of F. Galton (1879).

[3]The account of *psychodynamic process* entails interaction of common sense entities at a behavioral level. A *process theory* entails the continuous flux of abstract entities at a microtemporal level, where the stability of mental objects is not a given but an achievement.

[4]Freud postulated an intrinsic and an extrinsic source of Q.

[5]Others have come to similar conclusions, that Freud 'took the basically association-istic, mechanistic model of mind . . . and introduced a dynamic component' (Sirkin and Fleming, 1982).

[6]To say that 'a stimulus endures is not to say the sensation is sensed as enduring but only that the sensation also endures. The duration of sensation and the sensation of duration are different' (Husserl, E., 1905–1964).

[7]Whitehead argued that the lapse of time before a perception due to the finite velocity of luminal transmission (that is, that perception is never exactly *on-line*) implied that perception was always of the past, with memory being the basic paradigm for aware-ness. The report of a perception accentuates the lag. This is true for both perception and dream. In a Mahayana sutra, it is written:

> Immediate awareness is the same as in dreams, etc.
> At the time when immediate awareness has arisen,
> Seeing and its objects are already non-existent;
> How can it be admitted that perception exists?

The problem of a delay before a perception in relation to memory and awareness is discussed in Self and Process (Brown, 1991), as is the related fact that the report-state for a dream is subsequent to the actual dream state so that one never has a direct account of the dream image (that is, we do not know whether the dream we recall is the actual dream we experience). In this regard see Dreaming (Malcolm, 1959). Similarly, the delay before an action is ingredient in the feeling of agency (Brown, 1996, p. 115 et seq.)

[8]Thought and desire go out to objects and are intentional, while wish fulfillment in dream involves concrete images without an accompanying desire. In dream, self and world are incompletely individuated; time and intentional feeling are collapsed in a single fluid dream-space.

[9]This is an old complaint. Bradley (1897) wrote of association theory that the 'associ-ated elements are divorced from their temporal context; they are set free in union and ready to form fresh unions without regard for time's reality.'

[10]Consistent with E. von Hartmann (1893).

[11]See the section on hallucination and imagery in Life of the Mind (Brown, 1988, pp. 206–251).

[12]Brentano thought that 'in the sphere of phantasy we have uncovered the origins of ideas of time . . . [in the persistence of fading memories of events] . . . which are shoved back more and more.' Köhler (1923) held a similar view. See Self and Process (Brown, 1991, pp. 127–146) and Time, Will, and Mental Process (Brown, 1996. pp. 17–57 et seq.) for an account of subjective time in a microgenetic framework.

[13]The emergence of the self at pre-processing stages is in accord with the concept that evolutionary advance is by way of subsurface branching, not terminal addition.

[14]The distinction of a core self (matrix, essential, immutable) and conscious self (accidental, peripheral, changing) is an old one (Frondizi, 1955; Parker. 1941). The core self accounts for continuity, the occurrent self for variability, and the postulation

of two levels in the self resolves the impasse between those for whom the self is a collection of accidental features – Plato referred to the self as a society, Hume as a commonwealth – and those for whom continuity is primary. The resolution comes about as the persistence *distributes* into momentary instantiations, while the instantaneity expresses momentary features of an enduring potential at the core.

[15]The core self is associated with experiential cognition, the conscious self with explicit concepts. This is similar to the distinction of non-verbal experiential knowledge of the self and a self-concept or self knowledge that is language-dependent (Evans, 1970).

[16]On the distinction of subject and self, see Time, Will and Mental Process (Brown, 1996, pp. 1–21, 169).

[17]Freud relied heavily on the evidence from 'parapraxes' and other symptoms in pathological cases for the concept of *Cs* and *Ucs* cognition.

[18]If one applies to *Cs* the Aristotelian distinction of a thing and its attributes, the description of the attributes of *Cs* entails that *Cs* is a substance or entity which these attributes modify. Whitehead attributed the subject/predicate logic of substantialist thought to the structure of classical Greek. I would suspect that grammar expresses and reinforces trends of thought determined by the patterns of object perception.

[19]Locke maintained that if a sleeping person thinks without knowing it, he is a different person than when awake since *Cs* constitutes personal identity. However, there is an identity of-the-occasion and there is an identity as character or personality. These two sources of identity are comparable to the core and the *Cs* self. Identity in the sense of character is bound up with traits, values and beliefs that appear to be largely *Ucs* though we often judge a person's character after a brief encounter. It may be that as character develops it conforms to an average of the momentary instantiations of the core self which, over time, would exhaust the potential of the self and so reveal the range of possibilities that a given character permits.

[20]On the continuum from dream to wakefulness, see: Aristotle (1931). On transitional states and sleep-talking, see Brown (1988, p. 215).

[21]Compare with E. von Hartmann (op. cit., p. 49), 'it follows that now and never can consciousness *frame a direct conception* of the mode and manner in which the unconscious idea is presented'; with Ribot, 'subconscious activity is purely cerebral; the psychic factor which ordinarily accompanies the work of the nervous centres is absent' (Ribot, 1910); and with Freud: 'How are we to arrive at a knowledge of the unconscious? It is of course only as something conscious that we know it, after it has undergone transformation or translation into something conscious' (SE 14, p. 166).

The presumed inability to describe the *Ucs* except for those elements that become *Cs*, i.e., the lack of *Cs* access to *Ucs* events *qua Ucs*, thus the equating of the *Ucs* with the physical, runs parallel to the lack of *Cs* access to the physical brain and world process that underlies *Cs* experience. One can agree with Whitehead (in Process and Reality) that the universe consists of elements disclosed in the experiences of subjects, and that apart from such 'drops of experience . . . there is nothing, nothing, bare nothingness.'

[22]In my view it is a waste of time to address the errors of minor philosophers who continue to deny the primacy of the subjective yet do not account for what on their position is the chief illusion of existence (See Evans [1970] for a critique of the 'persons-approach', i.e., the elimination of the immediate, private data of *Cs* by restricting its definition to what is ascribed by an observer to another person.)

[23]Piaget distinguished an early stage of awareness of objects and activity in children from a later stage of self-awareness. One could say that a state of non-intentional subjectivity in which a *Cs* 'aboutness' relation has not been established anticipates and underlies one of a *Cs* and intentional self.

Notes to Chapter 5

[1] For Rapaport, 'all psychological phenomena originate in innate givens, which mature according to an epigenetic plan' (R: 809).

[2] A more general issue not addressed in this chapter concerns the uncertain import of an analysis of the *content* of a cognition for the process through which content actualizes. In microgenesis, content is not instrumental in process but perishes when process terminates.

[3] The problem of inter-subjective validation was revisited by: Grunbaum, A. (1984) The Foundations of Psychoanalysis. University of California Press, Berkeley.

[4] Again, see Hartmann for the complete inventory, which incorporates most of Freud's list, including instinct, dream, timelessness, certainty and drive-satisfaction.

[5] The doctrine of an association between mental solids follows from Freud's emphasis on a causal account of the mental. Freud wrote, 'the psychical apparatus must be constructed like a reflex apparatus' (SE 5:538), and 'we will picture the mental apparatus as a compound instrument, to the components of which we will give the name of "agencies", or . . . "systems"' (SE 5:536). The distinction between an agency and its function, or a structure and its activity, so prominent in Freud's writings, is similar to that between a centre and its output in the older aphasia models.

[6] Pribram, K. (1991) Brain and Perception. Erlbaum, New Jersey.

[7] Surprising, in view of Hughlings Jackson's descriptions of epileptic auras in which, it is now generally accepted, the hallucination is elicited as part of a 'dreamy' state.

[8] For Hartmann (PU 1: 306), the sense of 'former knowledge' does not lie in the idea, which appears as something new. Instead, the mark of pastness derives from a comparison of vivid and weak ideas. His discussion of fading sense-impressions in relation to occurrent ones has similarities with that of Köhler (Köhler, W. (1923) Zur Theories des Sukzessivverbleichs und der Zeitfehler. *Psychologische Forschung.* 4: 115–175) and my account of subjective duration and the incrementation of the present (Brown, J.W. (1996) Time, Will and Mental Process. Plenum, New York).

[9] Discussed in Brown, J.W. (1991) Self and Process. Springer, New York, 102.

[10] The citation is from 1894. The reticular theory of Golgi, which dominated neurology until the period 1888 to 1892, did not greatly influence Freud, though he did attempt to resolve the competing theories in the *Project*. The theory was challenged by the work of His, Forel and Cajal (see: Cajal, R. (1954) Neuron Theory or Reticular Theory. Madrid). Freud's association doctrine is a metaphor of synaptic transmission, which informed his early thought, persisted throughout his writings and was extrapolated to psychic function. After discussing the excitatory axo-dendritic contacts of neurons 'through the medium of a foreign substance', Freud wrote, 'a single neuron is thus a model of the whole nervous system' (SE 1: 298). (See also: Amacher, P. (1965) Freud's neurological education and its influence on psychoanalytic theory, Psychological Issues 4 (4), International Universities Press, New York). Thus, Strachey described the shift in Freud's thought from the *Project* to The Interpretation of Dreams as follows: 'The systems of neurones were replaced by *psychical* systems or agencies; a hypothetical 'cathexis' of psychical energy took the place of the physical 'quantity' (SE 4: xviii).

[11] Brown, J.W. (1997) Introduction to: Solms, M. The Neuropsychology of Dreams. Erlbaum, New Jersey.

[12] Discussed in Brown (1996).

[13] For the relation to earlier studies in aphasia, see: Greenberg, V. (1997) Freud and his Aphasia Book: Language and the Sources of Psychoanalysis. Cornell University Press, Ithica, and my review in: J. Nerv. Ment. Dis., 186: 383–385.

[14] The distinction is reminiscent of Wernicke's two paths in language – the automatic and the voluntary – which Freud criticized.

[15]The specification of meaning is linked to awareness, but specification also occurs in all other domains of cognition – phonology, perceptual analysis, etc. – where it is not tightly related to awareness. This implies that the *process* of specification is distinct from the generation of consciousness, which is related to the depth of the phase, especially those phases concerned with meaning.

[16]Freud asked if 'it is not the repression itself which produces substitutive formations and symptoms'? He concluded that the symptom was due not to the repression but to 'the return of the repressed' (SE 14: 154). Substitutions, displacements, hallucinations, neurotic behaviors, etc., are outcomes of the repression of unpleasant ideas, i.e. the symptom is not 'repression itself', but its result.

[17]Sander, F. (1928) Experimentelle Ergebnisse der Gestaltpsychologie. Berichte uber den X Kongress der experimentellen Psychologie im Bonn *1927*. See: Smith, G. and Danielsson, A. (1982) Anxiety and Defensive Strategies in Childhood and Adolescence. Psychol. Issues. 52, International Universities Press, New York.

[18]See Freud, SE 13.

[19]Brown, J.W. (1988).

[20]This argument was advanced in the earliest publications on microgenetic theory prior to work on blindsight, and 20 years before anatomical evidence of more prominent neocortical connectivity in a direction *from* V–4 *to* V–1 than in the opposite (standard) direction.

[21]The priority of an experience is not a simple matter of when the experience occurs, but reflects the pastness or presentness 'assigned' to that experience in subjective time. See: Brown, J.W. (1996).

Notes to Chapter 6

[1]*Saptaśatikā*: 213.

[2]See Conze, E. (1968) Thirty Years of Buddhist Studies. University of South Carolina Press, Columbia. Also, Raju, P. (1953) Idealistic Thought of India. Harvard University Press, Cambridge.

[3]On the role of negation in Indian philosophy, see: Upadyaya, K. (1988) Indian tradition and negation. Philosophy East and West 38: 281–289. Upadyaya considers negation a cognitive derivative of a positive situation in reality. Whitehead, writing on the discriminative powers, remarks that 'the negative judgment is the peak of mentality'.

[4]Chang, G.C.C. (1983) A Treasury of Mahāyāna Sūtras. Penn State University Press, Pennsylvania (Edition: Banarsidass, 1991, p. 172).

[5]See: Whitehead, A.N. (1938) Modes of Thought. Macmillan, New York.

[6]For other examples, see: Carey, S. and Gelman, R. (1991) Eds. The Epigenesis of Mind: Essays on Biology and Cognition. Erlbaum, New Jersey.

[7]Whitehead, A.N. (1934) Nature and Life. Cambridge University Press, Cambridge.

[8]Matilal, B. (1986) Perception: An Essay on Classical Indian Theories of Knowledge. Oxford University Press, Oxford.

[9]McDowell, J. (1994) Mind and World. Harvard University Press, Cambridge.

[10]Arguments for this theory of duration and its details are discussed in: Brown, J.W. (1993) Self and Process. Springer-Verlag, New York, and Brown, J.W. (1996) Time, Will and Mental Process. Plenum, New York.

[11]The uncoupling of the subjective present from physical passage is a fundamental, if ignored, obstacle to mind/brain reduction. Paul Weiss has written, 'The time of the mental event is a stretch of which the time of the physical event is a moment is an

hypothesis that seems never to have had the consideration it deserves.' (Weiss, P. (1936) Philosophical Essays for Alfred North Whitehead. Longmans, Green, London, p. 156). According to Carnap, 'Einstein said that the problem of the Now worried him seriously . . . (and the fact that) this experience cannot be grasped by science seemed to him a matter of painful but inevitable resignation'. (Carnap, R. (1963) Autobiography. In: P. Schilpp (Ed) The Philosophy of Rudolf Carnap. Open Court, La Salle.)

[12]Kalupahana, D. (1992) A History of Buddhist Philosophy. University of Hawaii Press, Hawaii.

[13]The pure land as a popular expression of paradise is an objectless state of limitless duration. In this respect, it is comparable to a state of pure consciousness in an act of cognition. Avatamsaka has written: 'All the Buddha-lands rise from one's own mind and have infinite forms' (Conze, E. (1951) Buddhism. Bruno Cassirer, Oxford, p. 156).

[14]Renau, L. (1950) La civilisation de l'Inde ancienne d'après les textes sanscrits. Paris; Paz, O. (1995) In Light of India. Harcourt Brace, New York.

[15]Coomaraswamy, A. (ed: 1993) Time and Eternity. Munshiram Manoharlal, New Delhi, p.47.

[16]Akira, H. (1990) A History of Indian Buddhism. University of Hawaii Press, Hawaii.

[17]Inge, W. R. (1929) The Philosophy of Plotinus. Longmans, Green, London.

Notes to Chapter 7

[1]Arthur Danto has written, with irony, that the pure self is needed to remind ourselves of the potential to have been another self had the circumstances that determined the empirical self been different.

[2]The most recent proposal, a scanning or binding device that synchronizes the elements, is an ad hoc integration of units that were needlessly stripped of their temporal relations in the first place.

[3]Some cases have a visual experience at a preconscious level (for the original description, see Bender and Krieger, 1951).

[4]There stands a man, staring upward, wringing his hands in the grip of pain. It sickens me when I behold his face, the moon reveals my own shape.

[5]Brown J and Grober E (1982). Unpublished research results.

[6]In this figure from James' discussion, A, B, and C stand for three successive thoughts, each with its object inside of it. Köhler wrote of the comparison of A and B in relation to subjective time and the feeling of pastness. In my version, in the middle figure, the decay point of A (or the extent to which it is revived in B) represents the posterior boundary of the duration of the present; the surface of B represents the occurrent actuality. The duration of the present is extracted from the discrepancy of B and A. The object world is realized at B. The self is 'situated' at the point where A fades into the unconscious of 'long-term memory' (Brown, 1996).

[7]For a discussion of the problem of coherence in dream and in wakefulness see Malcolm (1959).

Notes to Chapter 8

[1]Collingwood, R. (1946) The Idea of History. Clarendon, Oxford.

[2]Whitehead, A.N. (1926) Modes of Thought. Macmillan, New York. Whitehead's insight is nicely illustrated in a remark by Marvin Minsky, 'You can think of the brain as a big supermarket of goodies that you can use for different purposes' (NY Times July 28, 1998).

[3]For a discussion of quantum mechanics in relation to emergence and consciousness, see: Silberstein, M. (1998) Emergence and the mind-body problem. Journal of Consciousness Studies 5: 464–482.

[4]Danto, A. (1997) After the End of Art. Princeton, New Jersey, p. 35.

[5]Discussed in: Brown, J.W. (1996) Time, Will and Mental Process. Plenum, New York, p. 239 et seq.

[6]Recall the automatic sweetheart of William James, the soulless beauty who displays love but does not feel it, in which the relation of mind to brain was analogized to that of god to nature.

[7]Heidegger, M. (1953) An Introduction to Metaphysics. Yale University Press, New Haven.

[8]Since scientific paradigms, according to Kuhn, use incommensurable concepts and methods, Kim asked why not accept them all without fear of logical incoherence, that is, if there is no principle of explanatory exclusion, one could have scientific progress through an accumulation of paradigms (Kim, J. (1989) Mechanism, purpose and explanatory exclusion. In: J. Tomberlin (Ed.) Philosophical Perspectives, 3, Philosophy of Mind and Action Theory. Ridgeview, California).

[9]Durckheim, E. and Mauss, M. (1903) De quelques formes primitives de classification. L'Année sociologique 6: 1–72. Lévy-Bruhl, L. (1910–1985) How Natives Think. Engl. transl. by L. Clare, Princeton, New Jersey.

[10]For Emmet (Whitehead, Heidegger, Cassirer), 'something like a "pre-animistic" stage underlies experience'. This phase was conceived as a relational ground from which self and world individuate. See: Emmet, D. (1945) The Nature of Metaphysical Thinking. Macmillan, London.

[11]Evidence for this is in: Brown, J. W. (198) The Life of the Mind. Erlbaum, New Jersey.

[12]Whitehead, A.N. (1926) Religion in the Making. Macmillan, p. 23

[13]An example might be the similarity across very different cultures of the sand paintings of the Navaho to Buddhist mandalas with respect to a complex imagery of body and cosmos.

[14]For an unintentional parody of reductionist logic, see: Hobson, J. (1988) The Dreaming Brain. Basic Books, New York.

[15]Levi-Strauss, C (1964) Le Cru et le Cuit. Plon, Paris. The binary system of Jakobson, which was the basis for this classificatory scheme of myth, was applied by Luria to neuropsychology with even less success.

[16]Wittgenstein compared the myth to Freudian dream analysis, writing that 'Freud has not given a scientific explanation of the ancient myth. What he has done is to propound a new myth'. In: Wollheim, R. and Hopkins, J. (1982) (Eds) Philosophical Essays on Freud. Cambridge University Press, Cambridge, p. 9.

[17]Lévy-Bruhl notes that participation is a holistic enterprise with a prominence of affective elements that does not involve external relations between the participating objects (Lévy-Bruhl, L. (1949–1975) The Notebooks on Primitive Mentality, Harper and Row, New York). The terms of the participatory complex are unindividuated in a primitive conceptual whole. This agrees with accounts of preliminary cognition as a presentation of incompletely analyzed formations accompanied by intense affects.

[18]For example, the formation of symptoms in hysteria, or the common onset of delusions of extra-terrestrial visitations or 'body-snatching' in hypnotic states.

[19]Aphasics sort objects according to functional similarities rather than abstract class. For example, they match a glove with a hand, not a hat, ignoring the abstract grouping for the experiential one, or matching a tiger with a crocodile, not a cat, on the basis of ferocity.

[20]Compare with psychotic cases: I am a virgin/Mary is a virgin/I am the virgin Mary. A fragment of paralogical or metaphoric thinking is no doubt responsible for empathy and the identification that is essential to the experience of art.

[21]The Trumai are real men, not reflections. Can the perception of a reflection, where the image of a thing is perceived as distinct from the thing itself, and the presence of an image supposes knowledge of the real, be read as a metaphor for the beginnings of introspection? Compare with the discussion of the myth of Narcissus in: Danto, A. (1981) The Transfiguration of the Commonplace. Harvard University Press, Cambridge.

[22]Lakoff, G. (1987) Women, Fire and Dangerous Things. University of Chicago Press, Chicago.

[23]Perhaps owing to, or reflecting, the deletion (enthymeme or lack of awareness?) of the connecting term.

[24]Lévy-Bruhl, L. (1935) Primitive Mythology. Engl. transl. 1983, University of Queensland Press.

[25]Elkin, A. (1943) The Australian Aborigines. Angus and Robertson, Sydney.

[26]To the primitive mind, nature spirits like the ghosts of the dead are largely pernicious. The persecutory character may be compared to that in paranoia, and interpreted as a carry-over into wakefulness of a fragment of dream cognition in which the agentless self is a victim of its own imagery.

[27]In Frazer, J.G. (1927) Man, God and Immortality. Macmillan, New York. For Freud, the tribal taboo arises from 'a temptation in phantasy set in motion though the agency of unconscious connecting links' (Freud, S. (1913) Totem and Taboo. Standard Edition 13: 1–18, Hogarth Press, London, 1955).

[28]Compare with the German superstition not to leave a knife with the edge upwards for it might injure God or an angel.

[29]This example of consubstantiality is a common feature of mythic consciousness, for example, the belief that an individual may be both dead and alive or inhabit two different forms at the same time. The European idea of a soul that detaches from the body at death is radically different from the magical idea of soul and body as an indissoluble unity but there are vestiges, in the Eucharist communion, of the tribal belief that eating the body of the deceased transmits character, as in cannibalism, where one acquires the qualities of a person by eating his body parts. In primitive thought, the person is a psychic entity that continues unembodied after death. In historic Uganda, the wives of a departed king are not 'widows' but act as if he is invisibly present (Lévy-Bruhl, The Soul of the Primitive, p. 236).

[30]In magical practise, or in voodoo, it is a common belief that one can harm an individual by a hostile act toward a fragment of the body, the hair, nails or clothing, or even a possession. The dead can return as malicious ghosts if the property left behind is not accorded a proper respect. A spear impaled on a footprint impales the man or animal who produced it. Lévy-Bruhl notes the ubiquity of *pars pro toto as* an immediate datum in the sense that the whole is felt to be present in the part (Notebooks, p. 84).

[31]In this respect, one can agree with many contemporary anthropologists who, like Boas, maintain that 'it would be an error to suppose that these (superstitious) beliefs are confined to the uneducated (rather) such belief may persist as an emotionally charged tradition among those enjoying the best of intellectual training. Their existence does not set off the mental processes of primitive man from those of civilized man' (Boas, F. (1938) The Mind of Primitive Man. Macmillan, New York).

[32]Notebooks, p. 58.

[33]Malin writes: 'The standard view of the universe as inanimate is a characteristic of the scientific method, not of the universe'. He goes on to compare elementary quantum events to the objectivized actual entities of process philosophy (Malin, S. (1997) Delayed-choice experiments and the concept of time in quantum mechanics. In: H. Atmanspacher and E. Ruhnau (Eds) Time, Temporality, Now. Springer, Berlin, pp. 43–52). For a quantum model of mind that has similarities with microgenesis and process philosophy, see: Stapp, H. (1993) Mind, Matter and Quantum Mechanics. Springer-Verlag, Berlin.

[34]Williams, B. (1976) Moral luck. Proceedings of the Aristotelian Society, suppl. 1: 115–135.

[35]See the discussion in: Tversky, B. and Hemenway, K. (1984) Objects, parts, and categories. Journal of Experimental Psychology 113: 169–193.

[36]A complex topic, discussed in: Brown, J.W. (1996) Time, Will and Mental Process. Plenum, New York.

[37]The discovery by Descartes of an interior space of psychic entities justified by the capacity for doubt coincides with the notion of introspection as an exuberance of pre-selection before final definiteness.

[38]Churchland, P.S. (1996) The hornswoggle problem. Journal of Consciousness Studies 5–6: 402–408.

Notes to Chapter 9

[1]Danto, A. (1981) The Transfiguration of the Commonplace. Harvard University Press, Cambridge. p. 148.

[2]Williams, C. (1998) Modern art theories. Journal of Aesthetics and Art Criticism 56: 377–389.

[3]Plotinus (1921 ed.) Psychic and Physical Treatises. Vol II, Translation by S. Mackenna of the Second Ennead; ca 260AD; Warner, London.

[4]For a recent review of philosophical writings, see: Rescher, N. (1996) Process Metaphysics, SUNY Press, New York.

[5]Brown, J.W. (1988) Life of the Mind. Erlbaum, New Jersey.

[6]Zaimov, K., Kitov, D. and Kolev, N. (1969) Aphasie chez un peintre. Encephale 58: 377–417; see also: Leischner, A. (1979) Aphasien und Sprachentwicklungsstorungen. Thieme, Stuttgart. p. 314. These cases and others are described in Brown, J.W. (1977) Mind, Brain and Consciousness. Academic, New York. The topic has recently been discussed from the standpoint of hemispheric interaction by Nikolaenko, N. (1998) Brain Pictures. Osaka.

[7]The possibility of a 'picture-strip' memory was discussed by McCulloch, W. (1965) Embodiments of Mind. MIT Press, Cambridge.

[8]For example, in déjà vu, the feeling of familiarity for novel events indicates that recognition and familiarity are not wholly driven by recurrence.

[9]McGinn, C. (1982) A note on functionalism and function. In: J. Biro and R Shahan (Eds) Mind, Brain and Function. University of Oklahoma Press, Norman. pp. 169–170.

[10]Whitehead, A.N. (1929) Process and Reality. Cambridge University Press, Cambridge. p. 427.

[11]Though mathematicians conceive non-Euclidean geometry without visualizing it. See Reichenbach, H. (1957) The Philosophy of Space and Time. Dover, New York. p. 48.

[12]See Alperson, P. (1980) 'Musical time' and music as an 'art of time'. Journal of Aesthetics and Art Criticism 38: 407–417.

[13]The perception of a melody with the eyes closed, and the progressive elimination of secondary qualities, was used as an illustration by Bergson of pure duration. For example, Bergson, H. (1923) Durée et Simultanéité. Felix Alcan, Paris. p. 55

[14]Albersheim, G. (1964) Mind and matter in music. Journal of Aesthetics and Art Criticism 22: 289–294.

[15]On the contrast of clock or external time with music as virtual time, see Langer, de Selincourt.

[16]Morel, F. (1936) Des bruits d'oreille, des bourdonnements, des hallucinations auditives élémentaires, communes et verbale. Encephale 31: 81–95. See also, Brown, J.W. (1988) The Life of the Mind. Erlbaum, New Jersey. p. 216.

[17]See the account by Nabokov in his autobiography, Speak Memory.

[18]A conclusion shared by Cytowic, R. (1989) Synesthesia: a Union of the Senses. Springer-Verlag, New York.

[19]Consistent with this view, Turner writes, 'periodicity, interval, motion, and closure belong to the whole of experience and are represented by music, painting, sculpture, poetry, theater, and dance equally.' Acknowledging the many differences between them, he goes on to say that parallels can be reduced 'to a common denominator, the universe of human feeling.' Turner, N. (1998) Cezanne, Wagner, Modulation. Journal of Aesthetics and Art Criticism 56: 353–364.

[20]From the Ode on a Grecian Urn: 'Heard melodies are sweet, but those unheard/ Are sweeter'.

[21]Whitehead would refer to the 'subjective aim' in the concrescence of a non-cognitive object.

[22]On an initial stage in thought of diffuse subjectivity continuous from inner to outer, see: Durckheim, E. and Mauss, M. (1903) De quelques formes primitives de classifications. L'Année Sociologique 6: 1–72; on the 'primitive' sense of community as a whole entity rather than a collection of parts, see: Lévy-Bruhl, L. 1975) The Notebooks on Primitive Mentality. Harper & Row, New York.

[23]The Aristotelian definition of art as a skill in making (poesis) rather than an action flowing from character (praxis) objectifies art by fixing its aim in an object (Shusterman, R. (1995) In: S. Gablik (ed.) Conversations before the End of Time. Thames and Hudson, London. p. 254).

[24]See discussion in: Brown, J.W. (1996) Time, Will and Mental Process. Plenum, New York.

[25]Schafer, E. (1961) Aesthetic perception. In: I. Leclerc (ed.) The Relevance of Whitehead. George Allen and Unwin, London. For a discussion of the intrinsicness of aesthetic experience from the standpoint of process thought, see: Hall, D. (1973) The Civilization of Experience. Fordham University Press, New York.

[26]Cassirer, E. (1944) An Essay on Man. Yale University Press, New Haven.

[27]Collingwood, R. (1938) The Principles of Art. Clarendon, Oxford.

[28]On the role of ambiguity in art, see: Kris, E. (1952) Psychoanalytic Explorations in Art. New York; Gombrich, E. (1960) Art and Illusion. Phaidon Press, London.

[29]Gombrich: 'the old insight that it's naïve to demand that a painting should look real is gradually giving way to the conviction that it is naïve to believe any painting can ever look real.' Gombrich, E. (1960) Art and Illusion, Phaidon Press, London. p. 247. On the distinction of the pictorial and representational in music, see: Kivy, P. (1984) Sound and Semblance, Princeton University Press, Princeton.

[30]This, of course, is an old observation. Croce writes, 'if photography have anything in it of artistic, it will be to the extent that it transmits the intuition of the photographer'. Croce, B. (1909) Aesthetic. Macmillan, London. p. 28.

[31]Goethe titled his autobiography, Poetry and Truth, to indicate that some truths can only be conveyed poetically. Novalis said, 'the more poetic, the more true.' The differ-

ence is that of logic and authenticity, of a scientific truth and the truth of a genuinely real expression.

[32]Dewey, J. (1935) Art as Experience. Minton, Balch.

[33]Bosanquet, B. (1923) Three Chapters on the Nature of Mind. Macmillan, London.

[34]Vygotsky wrote that the source of artistic pleasure is 'the parasitic enjoyment of exploiting somebody else's labor free of charge' (Vygotsky, L. (1971) The Psychology of Art, MIT Press, Cambridge. p. 32). I would substitute symbiotic for parasitic. The fusion is more a becoming-one with the art than with the artist.

[35]See: Wordsworth's essay, The Sublime and the Beautiful.

[36]Kant, I. (1960/1764) Observations on the Feeling of the Beautiful and Sublime. trans. J. Goldthwait, University of California Press, Berkeley.

References

Aristotle (1931) De somnis. In The Works of Aristotle, Vol. III. Oxford: Oxford University Press.

Bender M, Krieger H (1951) Visual function in perimetrically blind fields. Archives of Neurology and Psychiatry 65: 72–99.

Binswanger L (1957) Sigmund Freud. Reminiscences of a Friendship. New York: Grune and Stratton.

Bohm D (1965) The Special Theory of Relativity. New York: Benjamin.

Bohm D (1980) Wholeness and the Implicate Order. London: Routledge and Kegan Paul.

Bradley F (1897) Appearance and Reality, 2nd edn. Oxford: Clarendon Press.

Bradley J (1994) Transcendentalism and speculative realism in Whitehead. Process Studies 23: 155–191.

Brentano F (1973) Psychology from an Empirical Standpoint. Ed. O Kraus (English translation of 1874 original). London: Routledge and Kegan Paul.

Brown JW (1988) Life of the Mind. Hillsdale, NJ: Erlbaum. [LM]

Brown JW (1991) Self and Process. New York: Springer-Verlag. [SP]

Brown JW (1994) Morphogenesis and mental process. Development and Psychopathology 6: 551–564.

Brown JW (1996) Time, Will and Mental Process. New York: Plenum. [TWMP]

Cajal R (1954) Neuron Theory or Reticular Theory. Madrid.

Deacon T (1989) In E. Perecman (ed.), Integrating Theory and Practice in Clinical Neuropsychology, pp. 1–47. Hillsdale, NJ: Erlbaum.

Dipert R (1997) The mathematical structure of the world. Journal of Philosophy 44: 329–358.

Evans C (1970) The Subject of Consciousness. London: George Allen & Unwin.

Frazer JG (1927) Man, God and Immortality. New York: Macmillan.

Freud S (1891/1953) On Aphasia. E. Stengel (Trans.). New York: International Universities Press.

Freud S (1953–1974) The Standard Edition of the Complete Works of Sigmund Freud. Ed. J Strachey. London: The Hogarth Press and The Institute of Psycho-Analysis. [SE]

Frondizi R (1955) The Nature of the Self. New Haven, CT: Yale University Press.

Galton F (1879) Psychometric experiments. Brain 2: 149–162.

Gill M (1963) Topography and Systems in Psychoanalytic Theory. Psychological Issues 10. New York: International Universities Press. [G]

Goldstein K (1939) The Organism, p. 310. New York: American Book Company.

Griffin D (1986) Time and the fallacy of misplaced concreteness. In D Griffin (ed.), Physics and the Ultimate Significance of Time, pp. 1–48 New York: SUNY Press.

Hartmann E von (1868–1931) Philosophy of the Unconscious, Vols 1, 2, 3. London: Kegan Paul, Trench, Trubner & Co. [PU]

Husserl E (1905–1964) The Phenomenology of Internal Time-consciousness. Bloomington: Indiana University Press.

Inge WR (ed.) (1929) The Philosophy of Plotinus (2 vols). London: Longmans Green.

James W (1890) Principles of Psychology. New York: Holt.

Klüver H (1933) The eidetic type. Association for Research in Nervous and Mental Disease 14: 150–168.

Köhler W (1923) Zur Theorie des Sukzessivvergleichs und der Zeitfehler. Psychologische Forschung 4: 115–175.

Langer S (1953) Feeling and Form. New York: Scribners.

Leclerc I (1958) Whitehead's Metaphysics. London: George Allen and Unwin.

MacIntyre A.(1958) The Unconscious. London: Routledge and Kegan Paul.

Malcolm N (1959) Dreaming. London: Routledge and Kegan Paul.

McTaggert J (1927) The Nature of Existence. Cambridge: Cambridge University Press.

Mill JS (1880) Gesammelte Werke, T. Gomperz (ed.), Vol. XII: Ueber Frauenemancipation, Plato. Arbeiterfrage Socialismus. S. Freud (Trans.). Leipzig: Fue's Verlag.

Parker D (1941) Experience and Substance. Ann Arbor, MI: University of Michigan Press.

Pribram K (1971) Languages of the Brain. Englewood Cliffs, NJ: Erlbaum.

Pribram K, Gill MM (1976) Freud's 'Project' Re-assessed. New York: Basic Books.

Rapaport D (1967) Collected Papers. Ed. M. Gill. New York: Basic Books. [R]

Rescher N (1992) The promise of process philosophy. In H Saatkamp, R Burch (eds), Frontiers in American Philosophy, Vol. 1. Texas: A & M Press.

Ribot T (1910) In T. Munsterberg et al., Subconscious Phenomena. Boston: R. Badger.

Schilder P (1953) Medical Psychology, p. 269. New York: International Universities Press.

Searle J (1992) The Rediscovery of the Mind. Cambridge, MA: MIT Press.

Searle J (1995) New York Review of Books, November 2, p. 6.

Selincourt B de (1958) Music and duration. In I Leclerc (ed.), The Relevance of Whitehead. London: George Allen and Unwin.

Seltzer B, Mesulam M (1988) Confusional states and delirium as disorders of attention. In F Boller, J Grafman (eds), Handbook of Neuropsychology, Vol. 1. Amsterdam: Elsevier.

Sirkin M, Fleming M (1982) Freud's 'Project' and its relationship to psychoanalytic theory. Journal of the History of the Behavioral Sciences 18: 230–241.

Sprigge T (1983) The Vindication of Absolute Idealism. Edinburgh: Edinburgh University Press.

Von Hartmann E (1893) Philosophy of the Unconscious, Vol. 2, 2nd edn, p. 78. London: Kegan Paul.

Wallack F Bradford (1980) The Epochal Nature of Process in Whitehead's Metaphysics. Albany, NY: State University of New York Press.[ENP]

Whitehead AN (1919) An Enquiry Concerning the Principles of Knowledge. Cambridge: Cambridge University Press.

Whitehead AN (1920) The Concept of Nature. Cambridge: Cambridge University Press.

Whitehead AN (1926) Religion in the Making. New York: Macmillan.

Whitehead AN (1929/1978) Process and Reality. Eds DR Griffin, D Sherburne. New York: The Free Press. [PR]

Whitehead AN (1948) Uniformity and contingency. In Essays in Science and Philosophy, pp. 100–111). London: Rider & Co.

Whitrow G (1961) The Natural Philosophy of Time. London: Thomas Nelson and Sons.
Wittgenstein L (1942–1982) In: Wollheim R, Hopkins J (eds), Philosophical Essays on
    Freud. Cambridge: Cambridge University Press.

Author Index

Printed in the United States
By Bookmasters